DISCOVERING BIRDS

DISCOVERING BIRDS

THE EMERGENCE
OF ORNITHOLOGY
AS A SCIENTIFIC
DISCIPLINE: 1760–1850

PAUL LAWRENCE FARBER

The Johns Hopkins University Press
Baltimore and London

Originally published as a hardcover edition in 1982 by D. Reidel Publishing Company.
Johns Hopkins Paperbacks edition, 1997
06 05 04 03 02 01 00 99 98 97   5 4 3 2 1

The Johns Hopkins University Press
2715 North Charles Street
Baltimore, Maryland 21218-4319
The Johns Hopkins Press Ltd., London

Library of Congress Cataloging–in–Publication Data

Farber, Paul Lawrence, 1944–
    [Emergence of ornithology as a scientific discipline, 1760–1850]
    Discovering birds : the emergence of ornithology as a scientific discipline, 1760–1850
    Paul Lawrence Farber.
        p.   cm.
    Originally published: The emergence of ornithology as a scientific discipline,
1760–1850. Dordrecht, Holland ; Boston, U.S.A. : D. Reidel, c1982, in series: Studies
in the history of modern science.
    Includes bibliographical references and index.
    ISBN 0-8018-5537-3 (pbk. : alk. paper)
    1. Ornithology—History.   I. Title.
QL673.F37 1997   96–41381
598´.09—dc20   CIP

A catalog record for this book is available from the British Library.

To the memory of
Fanny Sandler Schapiro,
Ruth Farber Robinson,
and Mickey Michaels

TABLE OF CONTENTS

PREFACE TO THE JOHNS HOPKINS EDITION

Until recently, much of the history of science has focused attention on the development of major scientific theories. Scholars examined the papers of famous figures such as Newton, Lavoisier, Darwin, Bernard, or Einstein, searching for the paths they followed in their quest to understand nature, or seeking the complex set of influences and intellectual interactions surrounding the origin and development of their theories. Historians investigating the social context of scientific theories added another, and highly valuable, facet to the story. They demonstrated the ideological purposes that science served, as well as the political, philosophical, and religious dimensions of scientific theories. Their studies have given us a richer and more fully textured account of science in its cultural context. In the last ten years, however, numerous historians have begun to question the central place that scientific theory occupies in the history of science, and many have suggested alternative perspectives on the scientific enterprise. They have successfully demonstrated the historical significance of laboratories, experiments, techniques, epistemic traditions, institutions, animal or plant research models, and equipment and have shown that much of this history can be pursued independently from the history of specific theories.[1]

Much of this work beyond scientific theory examines the methodology and the organization of research. Historians (aided by philosophers and sociologists of science) have fruitfully investigated the emergence of new disciplines, the transformation or merging of old disciplines, the existence of national scientific styles, and the methodological and philosophical assumptions of various research groups. These new, expanded studies are altering the perception we have of science, for they do not lend themselves to unified models of science (growth of

ix

empirical knowledge, paradigm shifts, etc.), as did the research on the origin and development of scientific theories. This should come as no surprise. The history of science now captures more of the multifaceted investigation of nature, which reflects the diversity of nature itself and the many possible perspectives on it. Natural history, for example, has had a long and complex history, which spans a period of time from Aristotle to the present. What was done under that heading, however, is enormously varied and has undergone major revisions.[2] Its development is not adequately conveyed by a conceptual history of the theories in the life sciences.

Among the most dramatic shifts in natural history was the breakup during the early nineteenth century of what for centuries had been a single subject into a set of related but distinct scientific disciplines. The story is not a subheading under the history of the theory of evolution's triumph, nor is it, as is often supposed, the chronicle of natural history's demise, with its theories of systematics being replaced by a broader, more inclusive "biology," one that constructs new theories of plant and animal function. Rather, what happened in natural history reflects a general trend toward specialization in nineteenth-century science, which has come to characterize modern investigations of nature. The result was that natural history ramified into a set of disciplines defined by objects of their study: birds (ornithology), fish (ichthyology), and so forth. The change came about because of a number of developments: the solution of technical problems, a vast increase in empirical information, the emergence of a set of institutions, the development of new audiences, the opportunities created by a wave of colonization related to the industrial revolution, the rise of new sources of funding, and the establishment of a critical mass of individuals studying specialized topics. The appearance of these new biological disciplines occurred in several countries at roughly the same time, and therefore their history avoids focusing on idiosyncrasies that might result from studying the development of disciplines in a single country.

This book is a case study of one of the first of the biological disciplines to emerge from the broader natural history. Although useful in understanding the origin of scientific disciplines, the book also carries wider value in demonstrating the complexity of science and scientific change. Many factors contributed to the emergence of ornithology, and only by looking at several of them – technical, empirical, social, cultural – can we begin to understand its history.

The history of science has had its share of disagreement over what constitutes the most fruitful method of investigation. When I entered graduate school in 1965, a controversy was under way between the "internal" and the "external" history of science. A wave of British-inspired social history soon overwhelmed that tired debate. More recently, various French and British versions of the social construction of knowledge have dominated much of the polemic. These arguments have honed the methodological sophistication of many practitioners of history of science, but they have given the misleading impression that one need choose a particular focus over others. This book illustrates the simple truth that we have much to gain from combining diverse perspectives. In natural history, the empirical, theoretical, social, and cultural are all critical dimensions that need to be studied in order to gain an understanding of its history.

NOTES

[1] A sample of some studies looking beyond theory: Peter Galison, *How Experiments End,* Chicago, University of Chicago Press, 1987; David Gooding, T. J. Pinch, and Simon Schaffer, (eds), *The Uses of Experiment: Studies in Natural Sciences,* Cambridge, Cambridge University Press, 1989; Robert Kohler, *Lords of the Fly: Drosophila Genetics and the Experimental Life,* Chicago, University of Chicago Press, 1994; Jane Maienschein, *Transforming Traditions in American Biology, 1880–1915,* Baltimore, Johns Hopkins University Press, 1991; Mary Jo Nye, *From Chemical Philosophy to Theoretical Chemistry: The Dynamics of Matter and Dynamics of Disciplines 1800–1950,* Berkeley, University of California Press, 1993; William Coleman and Frederic Holmes, (eds.), *The Investigative*

Enterprise: Experimental Physiology in Nineteenth-Century Medicine, Berkeley, University of California Press, 1988; Gerald Geison and Frederic Holmes, (eds.), *Research Schools, Osiris,* Vol. *8,* 1993; Albert Van Helden and Thomas Hankins, (eds.), *Instruments, Osiris,* Vol. *9,* 1994; Paul Lawrence Farber, "Theories for the Birds: an Inquiry into the Significance of the Theory of Evolution for the History of Systematics," in Margaret J. Osler and Paul Lawrence Farber, (eds.), *Religion, Science, and Worldview: Essays in Honor of Richard S. Westfall,* Cambridge, Cambridge University Press, 1985, pp. 325–339; and Nynn K. Nyhart, *Biology Takes Form: Animal Morphology and the German Universities, 1800–1900,* Chicago, University of Chicago Press, 1995.

[2] See for example N. Jardine, J. A. Secord, and E. C. Spary, (eds.), *Cultures of Natural History,* Cambridge, Cambridge University Press, 1996.

PREFACE

A number of years ago I began a project to define and evaluate the impact of Buffon's *Histoire naturelle* on the science of the late eighteenth and early nineteenth centuries. My attention, however, was soon diverted by the striking *difference* between the highly literary natural history of Buffon and the duller, but more rigorous, zoology of his successors, and I began to try to understand this transformation of natural history into a set of separate scientific disciplines (geology, botany, ornithology, entomology, ichthyology, etc.). Historical literature on the emergence of the biological sciences in the early nineteenth century is, unfortunately, scant.[1] Indeed the entire issue of the emergence of scientific disciplines in general is poorly documented. A recent collection of articles on the subject states:

One reason for this is, of course, that scientific development is a highly complex process. Consequently, there has been a tendency for those engaged in its empirical study to select for close attention one strand or a small number of strands from the complicated web of social and intellectual factors at work. Many historians, for example, have dealt primarily with the internal development of scientific knowledge within given fields of inquiry. Sociologists, in contrast, have tended to concentrate on the social processes associated with the activities of scientists; but at the same time they have largely ignored the intellectual content of science.[2]

The present study is a case study of the emergence of a particular scientific discipline. I have chosen to look at ornithology for it was among the first and the most important of the separate scientific disciplines to emerge in the early nineteenth century. The elements I have primarily focused on are those that I believe were central in this individual case. Only future studies on related disciplines will show how general the case was, however, given the influential role of ornithology in zoology this study can be expected to provide some elucidation on the emergence of nine-teenth-century biological disciplines.

My mother has often maintained that I was born under a lucky star, and although my scientific education has made me skeptical of the existence of such a causal agent, my life experience has provided an abundance of empirical evidence for her claim. I have been fortunate in a great number of ways; related to this project I have been especially favored in encountering helpful people and useful institutions. On a very basic level, I have been fortunate in obtaining a number of National Science Foundation grants, without which this project could not have been started, developed, or completed. I am very appreciative, therefore, for grants in 1973 (# GS 37955), 1975 (# SOC 75-14972), and 1977 (# SOC 77-26903). Oregon State University, through the Oregon State University Foundation and General Research Fund has generously supplemented the National Science Foundation grants by annually providing funds to help cover the expenses of photo-copying, microfilm, and domestic travel.

The National Science Foundation grants made it possible for me to work in a number of truly excellent libraries and institutions, in which I have consistently been helped by competent staff, most notably in the British Museum, British Museum (Natural History), Bibliothèque Centrale du Muséum National d'Histoire naturelle, Bibliothèque nationale, Library of Congress, Widener Library; the libraries of the Linnean Society of London, Royal Society, Museum of Comparative Zoology, University of Washington, University of California at Berkeley; the Archives de l'Académie des sciences, and the Archives nationales. Among the librarians and archivists who have helped me, and to whom I feel a great debt of gratitude, are Mrs. A. Datta, M. Y. Laissus, Mr. G. Bridson, and Mr. M. Kinch. Since 1968 Mme Taranne has been especially helpful to me in the everyday details of library use of the Muséum. Anyone who has had to work with a shaky command of the spoken language in a foreign library can well appreciate the value of a sympathetic staff member. One summer spent at Harvard was made more productive due to the gracious assistance of Everett Mendelsohn, and a winter's research in London was equally aided by the hospitality of Imperial College arranged by Rupert and Marie Boas Hall.

I also count myself lucky to have had a set of distinguished mentors, who have served as models to be emulated, if only imperfectly: Frederick B. Churchill and Richard S. Westfall, of Indiana University, provided the foundation of my education in the history of science; the late Joseph Schiller provided a "refresher" course in 1973 when I had the opportunity to see him regularly.

Colleagues, friends, students, and family have been supportive, and have substantially aided my work. Leona Nicholson, Karla Russell, and Dorothy Sheler provided skillful secretarial service all through the project. Keith Benson has not only patiently listened to hours of talk about the history of ornithology but has also kindly proofread this monograph. Tetty Schiller made our stay in Paris optimally productive and pleasant. Vreneli Farber helped me in all stages of the work, and provided an environment that made work on this study possible and enjoyable.

The following institutions have kindly granted me permission to reproduce archival materials: British Museum, British Museum (Natural History), Bibliothèque Centrale du Muséum National d'Histoire naturelle, Linnean Society, and the Archives nationales.

INTRODUCTION

The late eighteenth and early nineteenth centuries were a time of profound change in Europe. Historians have tried to convey a sense of the scope and magnitude of this transformation of European civilization by defining and describing a set of revolutions that took place in various domains: government, agriculture, commerce, manufacture, population growth, etc. Greatest attention has been focused on two events: the French Revolution of 1789, which marked a turning point in the political, social, and cultural predominance of the European aristocracy, and the British Industrial Revolution, which marked a starting point of an acceleration of economic expansion that influenced the lives of the entire population, not only in terms of an improved material existence, but also in such diverse areas as transportation, public health, organization of labor, and popular culture. Historians have also devoted careful attention to the agricultural revolution, the commercial revolution, the demographic revolution, etc., and the resulting vast body of historical literature on the period has led some researchers to attempt a synthesis and to construct a general picture of the transformation of European civilization. Unfortunately, one of the most striking developments in Western culture during the late eighteenth and early nineteenth centuries has received only minor attention in this latter endeavor. The fundamental alteration of the content, practice, and magnitude of the scientific enterprise is not adequately treated in general histories of the period.[1] To an extent, historians of science and technology have attempted recently to redress this neglect. They have investigated the major changes that took place in science and have attempted to relate those events to a broader historical picture.[2] This has been done most effectively by those who study the history of the physical sciences, such as chemistry, or those who study the social history of science. In so doing, these historians have revealed relationships that exist between the realm of abstract scientific ideas and the complex settings in which they are produced. They have conveyed the classic scientific texts from their holy of holies in the tabernacle of science and have brought them out to be viewed in their multifaceted historical splendor.

Historians of the biological sciences have been less active in attempting to integrate their studies on specific topics in this period into a more general historical account.[3] As a result, there exists less of an appreciation of the importance of the history of the biological sciences than the history of the physical sciences of the late eighteenth and early nineteenth centuries. In addition, interpretations in the history of the biological sciences have suffered because of a failure to go beyond intellectually narrow confines. This is most striking in the attempts to assess one of the central events in the history of the biological sciences: the transformation of natural history in the late eighteenth and early nineteenth centuries. That history was intimately connected with other cultural, economic, and political developments of the time. However, the two most widely held interpretations of the transformation of natural history both conceptualize too narrowly the factors involved and consequently miss much of the richness of their story.

One long-standing manner of describing the transformation of natural history has been to state that eighteenth-century natural history was superseded by nineteenth-century biology. Such an interpretation stresses the history of the method and of the subject of biology. Indeed, the word "biology" came into use in the early 1800's to describe what was then perceived to be a new approach to the study of the living world.[4] The polemical proponents of biology sought to distinguish that study, which was basically physiological, *i.e.*, a study of vital processes such as nutrition, respiration, and generation, from natural history which they regarded as primarily concerned with classification and description.[5] Gottfried Treviranus (1776–1837), one of the coiners of the neologism, wrote that biology would study "the different forms and phenomena of life, the conditions and laws under which they occur, and the causes by means of which they are brought into being".[6] Jean-Baptiste Lamarck (1744–1829), the other main coiner of the term, thought of biology as a new "theory of living organisms".[7] "Biology", *i.e.*, physiology, however, did not replace natural history, although physiology did become a fundamental science and a very exciting one. The general physiology of Claude Bernard (1813–1878), the cell theory of Matthias Schleiden (1804–1881) and Theodor Schwann (1810–1882), and the cellular pathology of Rudolf Virchow (1821–1902) stand as major milestones in the history of the modern life sciences and reflect the conceptual and methodological advances of nineteenth-century physiology. To evaluate the history of natural history, however, from the perspective either of the early polemicists, who hoped to establish a new science of biology, or of the later successes of their successors, is both unfair and a confusion of separate lines of development. Natural history had an independent evolution during the nineteenth century. Its significance, however, is not as universally recognized or appreciated by historians as is the development of physiology. The failure to evaluate adequately the history of natural history is in part a consequence of the influence of contemporary science on historical perception. Natural history today ranks quite low in the hierarchy of modern biological disciplines, whereas physiology is quite high. More important, but related, is the paucity of detailed studies written from a broadly historical perspective on the history of late eighteenth and early nineteenth-century natural history. Such studies would establish the magnitude of the enterprise of natural history, its institutional developments, its links with colonialization, its relationship with physiology and other biological sciences, and its popular appeal for reasons as diverse as natural theology and social legitimation.

Those few historians of science who have focused on the history of the natural history tradition have advanced another general interpretation of the transformation of natural history. They have, however, done so within the restricted confines of a view that is more philosophical than historical, and they depict the changes in natural history as a shift in the perception of nature from a descriptive and statically systematic one to a conception that included a temporal dimension: an intellectual shift from natural history to a history of nature. Arthur Lovejoy made famous the notion that a temporalization of the concept of nature took place during the eighteenth century,[8] and most of the current literature that stresses the shift from a static to a temporal view of nature is a reworking of — sometimes cast in a fashion quite at variance with — the Lovejoy theses.[9] As a high-level generalization in cultural history, the temporalization of the conception of nature may have value, but as an interpretation of the history of natural history it is unsatisfactory. Aside from being too Whiggish, in that it reorganizes the past in light of the "Darwinian Triumph", it is an inaccurate and inadequate interpretation. As late as the 1850's, the majority of the research in natural history remained as atemporal in conception as it had been in the 1750's. Unquestionably individuals from the mid-eighteenth century up through Darwin's time had made epistemological shifts that involved the employment of a temporal understanding of nature. There were, however, many other philosophical breaks during this period, and there were, as well, many other equally significant historical changes in science. The characterization of a century of work in terms of the novel philosophical assumptions of a few thinkers does not do justice to the wonderfully rich history of the study of nature's products.

What is needed to appreciate the transformation of natural history in the late eighteenth and early nineteenth centuries is a set of multifaceted studies which would elucidate and relate the main outlines of the story. A lot happened during this period. Natural history fragmented into several separate scientific disciplines. The quantity and quality of the material available for study increased dramatically. The criteria of what constituted serious science became more rigorous. Entirely new opportunities arose both for collecting material and for supporting those who wished to study it. The audience for natural history diversified and expanded. Significant debates were carried on in traditional areas such as systematics and nomenclature, and new tensions arose within disciplines, such as that between field and museum workers. As the entire enterprise of what had been natural history expanded, the differences among countries in their styles of research, institutional bases, philosophical assumptions, and sources of government support came into greater relief.

Such diverse elements when united into a history of the transformation of natural history are too broad to fit within the narrow strictures of either an approach that focuses completely on a history of conceptual and technical developments (the "internal" history of science) or one that depicts only the social and cultural context (the "external" history of science). Nor can the complexity of the story be conveyed by reference to general epistemological shifts, paradigm changes, or replacements of world view.[10] To understand this complicated episode in the evolution of European science, a synthetic approach is required. Careful attention must be paid to the detailed technical developments within natural history, to the philosophical assumptions that underlay those changes, to the history of the institutional settings in which the research took place, and to the social and cultural systems that supported, guided, informed, and reflected the evolution of natural history. The study needs to be comparative as well. Although the scientific community was international, national styles, philosophical assumptions, and opportunities for support differed significantly from country to country. An investigation of a particular national tradition, such as D. E. Allen's splendid book *The Naturalist in Britain*,[11] although extremely illuminating, does not allow us to differentiate the differences between national idiosyncrasies and the basic features of the evolution of natural history that transcend those particular contexts. Nor can the study of the history of natural history within one country permit us to appreciate the convergent lines of development that led to nineteenth-century natural history.

This monograph, *The Emergence of Ornithology as a Scientific Discipline: 1760–1850*, offers a case study of a central episode within the larger history of the transformation of natural history. Ornithology emerged as one of the first zoological disciplines during the fragmentation of natural history. It was a discipline that attracted considerable attention and support, and it was the setting for major theoretical debates as well as important empirical discoveries. For these reasons it can serve as an excellent example of the changes that took place in natural history during the late eighteenth and early nineteenth centuries. Furthermore, it raises questions and issues that need to be explored for a history of natural history that can be integrated into a general history of the period.

The history of ornithology, of course, is not virgin territory. It is a subject that has attracted keen interest for many years and from many different perspectives — mostly for iconography and bibliography in relation to systematics or art history.[12] Two general surveys of the history of ornithology have been written in this century: Maurice Boubier in 1925 published *L'Evolution de l'ornithologie*,[13] and Erwin Stresemann in 1951 published *Die*

Entwicklung der Ornithologie which has recently been translated into English with additional notes by G. William Cottrell.[14] Both Boubier and Stresemann were concerned with tracing what they regarded as the progressive development and accumulation of ornithological knowledge, and their perspective was from that of contemporary ornithology. Boubier and Stresemann therefore often evaluated ornithological work in terms of its present-day value rather than in terms of its historical significance. They also stressed those aspects of the past that have a continuity with modern ornithology. This study, in contrast to those whose focus is on iconography, bibliography, or recording the imperfect past, attempts to show how the study of birds went from being a neglected literary activity to become a scientific discipline that attracted a substantial group of scientists who shared a set of rigorous methods, exacting criteria, and ambitious goals. It also tries to relate these events to the conditions that made them possible; some of which were strictly empirical, some broadly philosophical, some as seemingly remote as the revolution in the printing industry or the expansion of the European colonial empires. The results, I hope, will bear on the more general question of the transformation of natural history in the late eighteenth and early nineteenth centuries. Only a full study of that chapter in the history of science will disclose to what extent this story is indeed representative. By attempting, however, to discuss in detail the many facets of the ornithological portion I hope to have provided a possible starting point of that wider issue.

DISCOVERING BIRDS

CHAPTER I

KNOWLEDGE OF BIRDS IN THE EIGHTEENTH CENTURY

If asked what was known about birds, an eighteenth-century gentleman would have responded by saying that a *great* deal was known about them. Birds occupied a major position in eighteenth-century culture. There was, in fact, no escaping them, for in seemingly every cultural area their presence was apparent.

In the exalted realm of the language of heraldry birds were the symbolic code of armorial insignia that designated the privileged few from the general mass of mankind. Eagles, pelicans, and martins were the most common avian forms on shields (escutcheons), but peacocks, doves, owls, and cocks, as well as mythical birds such as the phoenix, were depicted to signify position, relationship, or quality.

On a more mundane level, birds were an integral segment of the pleasures of the table. Lords as well as peasants were familiar with a wide variety of domestic and wild bird dishes. Gastronomes could have noted that some of the most famous gastronomic creations of the century were made of domestic fowl, such as *le poulet à la Reine* served to Louis XV's wife Marie Leczinska or *la perdrix en Chartreuse* created by Mauconseil, the cook to Louis XV's last mistress, Madame du Barry. These birds, in fact, constituted an important part of the economy and consequently were of interest to agronomists. Even savants, like Réaumur, considered the increase of poultry and egg production a serious and legitimate scientific endeavor. Numerous treatises, journal articles, and encyclopedia entries dealt with the practical aspects of the raising of pigeons, geese, ducks, and chickens for meat and eggs.[1] Game birds supplemented these domestic fowl, as evidenced by major

1

cookbooks which described as many as fifty wild birds suitable for cooking.[2]

Outside of their culinary value, game birds were noted for their place in the amusements of the eighteenth century. Hunting, either in the royal manner using falcons, or the ordinary manner using guns, was a central pastime of the period, and when not being eaten or chased, birds were thought to be entertaining in a decorative sense. During the eighteenth century a proliferation of bird keeping took place on a truly grand scale. In part, this was associated with the development of the English garden, which needed water fowl to complement its informal "natural" appearance. Aviaries housed birds that would not settle in the "natural" surrounding.[3] On a smaller scale, cage or chamber birds, which had charmed Europeans since the seventeenth century, continued to be an ever expanding commercial enterprise. Natives and exotics, such as parrots and cockatoos, were displayed in coffeehouses and fashionable drawing rooms, often in elegant bird cages, many of which survive from the period. Even dead birds, which due to improvements in fowling pieces and techniques of taxidermy could be artistically mounted, constituted an increasingly fashionable object to collect, and consequently made their appearance in the records of natural history auctions.[4]

The prevalence of avian motifs in much of the decorative arts underscores the position of birds in eighteenth-century culture. Porcelain, textiles, wallpapers, and pottery statuettes all reveal naturalistically depicted birds as a seemingly ever present motif. There were even mechanical bird toys; the most extravagant being Marie Antoinette's mechanical singing canary and Catherine the Great's peacock in a cage musical box.[5]

The world of fashion, that general barometer of cultural attitudes, also clearly reflects the importance of birds in eighteenth-century taste, for feathers and stuffed birds were among the most striking accouterments of the toilette. It was said that during

Joli Femme vêtu d'une Robe d'un nouveau gout dit a la Diane ; un ruban a la ceinture nouer en rosette sur le côté gauche, Coëffure surmonté d'un Pouf à l'Asiatique, orné d'une aigrette et d'une plume de Héron, d'un cordon de Perles, et d'un croissant de Diamants.

A Paris chés Basset Rüe S.t Jacques au coin de celle des Mathurins à l'Image S.te Genevieve. Avec Privilége du Roy

Illustration 1. "Joli Femme . . . orné d'une aigrette et d'une plume de Héron" from *Galleries des modes et Costumes Français*, Paris, Esnauts et Rapilly, 1778, Vol. 1, plate 18. (Phot. Bibliothèque nationale, Paris).

Marie Antoinette's time "when the Queen and the ladies of the court passed through the corridors of Versailles, one saw only a forest of plumes a foot and a half high playing freely above their heads".[6] The *folie des plumes* created by Léonard Autie, one of the most celebrated of the court hairdressers, has come to symbolize fashion in the *ancien régime*, and was a source of infinite imitation as well as caricature.

The more exalted realms of erudition and piety did not neglect feathered creatures. Birds were thought, for instance, to embody lessons for man. The abbé Noël-Antoine Pluche's *Le Spectacle de la nature* (1732-1750), which was among the most popular natural history treatises available at mid-century, described the spectacle of nature as a wondrous Creation replete with edification for man and intended to inspire a sense of awe before its All-Wise Creator. The intricacies of birds' nests, the variation and function of birds' beaks, the vast distances and complexity of birds' migrations were topics that illustrated the nature of Creation and its meaning for man.

Pluche's *Le Spectacle de la nature* was not a detailed scholarly treatment of birds. His audience consisted of young people and tyros, and he carefully avoided burdening his account with a surfeit of facts. Curiously, there was a paucity of natural history literature on birds. There existed a number of studies of local faunas from which one could cull fragments for a natural history of birds: Sir Hans Sloane, Francisco Hernandez, Gabrielem Rzaczynski, and Mark Catesby were among the most prominent authors. Some picture books were available, such as those done by Eleazer Albin, Johann Leonard Frisch, and George Edwards, and these served as a source for graphic design and iconographic reference. There were also a few taxonomic works; those of Jacob Klein, Pierre Barrière, Heinrich Möhring, and Carl Linnaeus were the most famous. These systems, however, were constructed with a superficial knowledge of birds, and the results were not very impressive. Considering how much was known about birds in various domains of knowledge, it is rather surprising how few scholarly works on birds existed. One could easily find a recipe to prepare a woodcock, a design for a neoclassical dovecote, or a reasonably accurate depiction of a cockatoo, but not a general natural history of birds until the works of Brisson (1760) and Buffon (1770), which will be discussed in the next chapter. All that was available before them were editions of Renaissance encyclopedists such as Pierre Belon, Conrad Gesner, and Ulisse Aldrovandi. These works, perhaps better illustrated and containing descriptions of more birds than editions of ancient encyclopedists like Pliny, were nonetheless the writings of a former age. To Enlightenment readers they must have appeared curious, incomplete, and unreliable. The Humanist tradition from which these works derived had, after all, a radically different perspective. It was quite legitimate, indeed important, for these encyclopedias to include articles on fabulous birds, such as the phoenix, and to dwell on philological and mythical details at length. By the mid-eighteenth century, an empirically oriented and experienced naturalist, like Réaumur, Buffon, or Pallas, could have read through the above mentioned authors and could have discovered many rich nuggets; however, he would have resented the amount of time spent sifting through what would have appeared as so much dross in order to separate the valid observations from the mass of accumulated folklore, naive descriptions, second-hand accounts, artists' imagination, and engravers' lack of skill. The most comprehensive and best ornithology that was available at mid-century, the posthumous ornithology of Francis Willughby edited by John Ray (1676), contained a classification system that was more satisfactory than contemporary or earlier and cruder attempts. Its illustrations, however, were notably poor and the number of species described was a mere five hundred. That Ray and Willughby's ornithology continued to be the most advanced ornithology for almost a hundred years after its publication says more about the continuity between the sixteenth- and seventeenth-century natural history traditions and the lack of innovation in the first half of the eighteenth century, than it does about the genius of its authors.

Considering the inappropriateness of Renaissance encyclopedias as a reference for Enlightenment readers, and keeping in mind the prevalence of birds in eighteenth-century culture, it is not surprising that in the 1760's a new sort of literature on birds appeared.[7] This literature was historically significant, not only because of its place in eighteenth-century culture, but also because it marked the start of a set of developments that led to the emergence of ornithology as a scientific discipline. This is not to say that other forms of the study of birds disappeared. Birds continued to play an important role in Western culture: in hunting, agronomy, cuisine, decorative arts, etc. What appeared in the third quarter of the eighteenth century, however, was the beginning of a new phenomenon that developed later to such a point as to make the claim that it encompassed "knowledge of birds", and that other domains, if they touched on birds, did so by merely applying this knowledge, or touched on the nature of birds in a secondary manner.

CHAPTER II

BRISSON AND BUFFON: ORNITHOLOGY 1760–1780

Two publications in the third quarter of the eighteenth century stand out as vanguards of a new type of study of birds: Mathurin-Jacques Brisson's (1723–1806) *Ornithologie* and Georges-Louis Leclerc de Buffon's (1707–1788) *Histoire naturelle des oiseaux.*

Illustration 2. "Brisson" engraving by G. L. Chrétien. (Phot. Bibliothèque nationale, Paris).

The first of these, Brisson's[1] *Ornithologie*, which was published in 1760, begins with what the author considered to be a history of ornithology up to his day, which is to say that he looks at the Renaissance encyclopedia tradition, starting his discussion with an appreciation of Belon and bringing it up through Ray. He then goes on to survey some more recent attempts to construct a classification system that would include all the birds known. According to Brisson all of these earlier works were inaccurate, too limited in scope and out of date. A lot more was or could be known about birds, and the time had come for a new ornithology to be written. What Brisson meant by more knowledge of birds was that considerably more species of birds could be listed than in any of the preceding treatises, including that of the great classifier of the century, Linnaeus, and that a new, coherent system of classification was needed to catalogue this expanded base of ornithology. Brisson felt that he was excellently situated to provide just such a study. Since 1749 he had been the *garde et démonstrateur* of René-Antoine Ferchault de Réaumur's (1683–1757) *cabinet d'histoire naturelle*, one of Europe's finest collections of natural history objects, and one that was especially rich in bird specimens.[2] The principal means of amassing a major collection in the eighteenth century was by means of an extensive network of correspondents. As "Prince of Naturalists" and as a member of the *Académie Royale des sciences*, Réaumur had an enormous scientific correspondence from which he could draw for specimens and/or descriptions. He had contacts all over Europe and, more important, he had correspondents in the colonies, who were able to supply him with specimens or descriptions of species wholly unknown to European naturalists. The value of Réaumur's collection was further enhanced by the care he took to obtain, when possible, more than one specimen per species, a sample nest, plus any information available on habitat and behavior. Réaumur also knew several travelers to the colonies or to other parts of the world, and these men were an additional source for his museum. Among the better known of Réaumur's correspondents were Pierre Poivre, Michel Adanson, Count Bentinck of Leyden, Jacques-François Artur, and Charles de Geer. Brisson was fortunate not only in having access to all of Réaumur's material, but also in being able to have the excitement of receiving along with Réaumur the new species, for many of the exotic bird specimens of Réaumur's museum were acquired just before or during Brisson's tenure as curator of the collection.[3]

We can better appreciate Brisson's position if we note that at mid-century birds formed a small part of most natural history collections. In part this was a consequence of fashion; shells were more the rage than stuffed birds. However, there was also a practical reason: the art of preserving bird specimens was very primitive. Réaumur, himself, noted the problem in a pamphlet he published on taxidermy, where he wrote:

That Part of Natural History which can offer to us the largest Series of agreeable Objects and actually offers a vast Number which are not sought after merely for the Pleasure of looking upon them; *viz.* that Part which treats of Birds, has remained as yet very imperfect, nor has it yet made them sufficiently known to us, because no considerable Collections have hitherto been made of them; and those who had begun to make any soon became weary of going on, having had the Mortification to see them every Day destroyed by ravenous Insects, in spite of all the care that had been taken to preserve them against their Teeth.[4]

Réaumur devoted considerable time and effort to the study of taxidermy, and although he did not solve its most serious problem − that of preserving specimens from insect attack − he did devise various methods of transporting and mounting birds. A few years before Brisson joined him, Réaumur had perfected a technique of drying birds in an oven in such a way that when done they presented a life-like appearance, and this stimulated him to assemble in a relatively short period of time the largest bird collection extant in Europe.[5]

Brisson, then, had the unusual opportunity of being placed in a private museum with an unparalleled bird collection that contained recently mounted birds in excellent condition and that included large numbers of exotic species, many unknown to science. Since Réaumur's residence and museum were on the outskirts of Paris, Brisson could also draw on the resources of the French capital where at this time natural history collections were very fashionable.[6] Although none could equal Réaumur's *cabinet d'histoire naturelle* in birds, Brisson profited from examining these other collections, and in his *Ornithologie* he cited over one hundred birds in the collections of the abbé Aubry, Madame la présidente de Bandeville, P. J. C. Mauduyt de la Varenne, Etienne-François, chevalier de Turgot, and the *Cabinet du Roi.*

Not only did Brisson benefit from the collections that were immediately accessible to him, but he also took his orientation from them. His ornithology has to be understood in the terms of a museum curator's relationship to a large private collection. Brisson reflected this orientation in the introduction of an earlier and more general work, *Le Regne animal divisé en IX classes:*

The position that I have had the good fortune to occupy for several years, which has put me in daily contact with the richest collection of nature's objects that has ever been made, has allowed me to make a great number of observations on the animal kingdom; to compare them and to examine the closest and most distant relations. I have been led to think of arranging the animal kingdom into an order different from those used up to the present time. My intention in this labor was solely to instruct myself and to place myself in the position of being able to judge the most convenient place to put a specimen of a new animal which would arrive to be placed in a cabinet.[7]

Brisson's approach to ornithology was *collection-catalogue natural history*, that is to say natural history written from a museum curator's perspective. We can see this in a number of aspects. On a superficial level it is evident where Brisson carefully told his reader which birds described were in Réaumur's collection, and where possible "the correspondent who has been willing to take the trouble to collect and send them to him".[8] (Approximately forty-five individuals were credited with sending three hundred and seventy-five different birds.) If the bird was not in the Réaumur museum Brisson identified the collection consulted.[9] In many ways, Brisson's *Ornithologie* was an expanded catalogue of the Réaumur collection. It described the nest and eggs of species when they existed in the collection, and it even recorded in an individual article "Le coq et la poule a cinq doits" – a monstrosity in Réaumur's cabinet. As might be expected with a collection-catalogue, the emphasis was on new species. For example, in his introduction Brisson stated that the two hundred and twenty plates, which he planned to publish to accompany the text, would illustrate roughly five hundred birds, three hundred and twenty of which had not been described before. This emphasis on new species, as well as Brisson's care to credit Réaumur and Réaumur's correspondents, his attention to Réaumur's nest collection and anomalies, all reflect in a fairly superficial way a collection-catalogue approach to natural history. But there was a deeper significance to his approach: to Brisson the study of birds was essentially a study of their arrangement, *i.e.,* their classification. Indeed, his most original work and his major contribution consisted of his new method of arranging birds, which was to place them in twenty-six orders defined by beaks and claws. This relatively large number of orders (compared to Linnaeus's six orders, for example) was a strength and a weakness of the system. Considering that the overall system was an artificial one, that is, it made no claims about the inherent relationships among the taxa and was constructed for convenience, the number of orders gave Brisson the flexibility to arrange birds in fairly natural groupings. To this day, ornithologists are impressed with Brisson's feel for bird groups.[10] The system, however, was awkward and difficult to remember, and in this sense the number of orders worked against Brisson in that it inhibited the spread of its acceptance on a wide scale. The details of his system faired better. Brisson described one hundred and fifteen genera in his twenty-six orders, and although a number of these genera were taken from Ray and earlier writers, sixty-five were new, of these sixty-four are still in use.[11] His entire system accommodated fifteen hundred species and varieties, three times as many as in Ray or in Linnaeus's tenth edition of the *Systema Naturae* published in 1758, two years before Brisson.

A careful reading of Brisson's *Ornithologie* shows the extent to which the entire six volumes is a collection-catalogue natural history, in the dual senses of being based on a particular collection and being primarily a classification. The overall organization is based on his twenty-six orders, which after an initial preface, are taken up one by one. Individual articles attempt to define species or varieties by lengthy descriptions of the external characteristics, often of particular specimens. Where he had seen a male, female, and juvenile, separate descriptions are made. The size of a bird is given as the size of a particular specimen, not as an average or typical size. Perhaps because Brisson was painfully aware of his limitations — often his specimens were in poor condition; he rarely had a male, female, and juvenile; he was wholly ignorant of the details of seasonal variation; and for a great many species he had no specimen at all — he compensated by producing, where he had adequate material, some of the finest descriptive ornithology in the literature.

Each article is in the same format: "First the size and proportions of the bird; next its colors, starting with the head and finishing with the tail."[12] Not the most exciting style for exposition, however, it was well suited for making comparisons. In like manner, the illustrations by François-Nicolas Martinet (b. 1731), which accompany the text, have the same stiff museum posture as Brisson's articles. These are valuable, however, in that they are basically accurate engravings, scaled for size, and they portray each genus. An extensive bibliography and discussion of nomenclature accompany each description. In part, this is a continuation of the Renaissance-encyclopedia tradition in which each article commences with a discussion of the history of the name. Brisson's treatment of nomenclature differs from the earlier treatments in that his emphasis is not so much on philology as on the practical necessity of knowing the history of the name in order for his readers to be clear on what bird it is that he is describing vis-à-vis the writings of other authors.

What Brisson omitted is as revealing of his intentions as that upon which he focused. His articles contain no information about a number of topics. Most notably, he has no discussion of the internal anatomy of birds — in general or specific. The lack of such a discussion is not surprising. Brisson was working with bird skins or mounted specimens. For this reason any discussion of the environment, distribution, or behavior, *i.e.*, any field information, is also absent. At best there is an occasional remark identifying the country from which the specimen was sent, or a comment on the bird's economic importance or its value as game.

The lack of detailed field information was a serious handicap for Brisson. It prevented him from giving an accurate account of the physical appearance of the life stages of different species, and the seasonal variations to which they are subject. It is to his credit that, at least, he was sensitive to and quite aware of the confusion generated by scant knowledge of life-stage histories, seasonal variations, sex differences, etc. In his discussion of birds of prey he lamented that to properly sort out their classification, one really needs to know the variation "at different times and during the entire course of their life, which is most often impractical".[13] His material occasionally forced him to describe a species based on a juvenile specimen or led him to describe as a new species the female of a known species. Given the material Brisson had at his disposal, it is hardly surprising that his classification system is based on simple external physical characteristics.

The *Ornithologie* consists of six volumes of a detailed bird classification, which was written in a singularly dull style, and limited its scope of investigation of particular species to an account of the bird's external physical appearance. For whom was Brisson writing? Clearly this was not a work intended for the general or casual reader, and this shows that by 1760 enough interest existed among collectors and amateurs to support the publication of an expensive six-volume work containing over two hundred engravings. The reception of Brisson's *Ornithologie* by this relatively small group was basically positive, and it was reprinted twice during his lifetime. Although his general system of twenty-six orders was never universally accepted, it was used by a number of naturalists and was considered to be the most important classification system until Temminck's work. The careful individual descriptions, his new genera, and the basically accurate plates by Martinet account for much of Brisson's contemporary reputation and lasting value.[14]

Brisson's *Ornithologie* is a good example of the collection-catalogue approach to natural history. The selection of material, the style, the scope, and the audience are all linked to a particular famous collection. That Brisson's ornithology was intimately tied to Réamur's museum can be further seen in the circumstances surrounding the termination of that research. Réamur, who died in 1757, had bequeathed his collection to the *Académie Royale des sciences*. On January 2, 1758, by royal ordinance, it was transferred to the *Cabinet du Roi* of the *Jardin du Roi,* a more appropriate setting for a natural history collection. Behind the transfer was Buffon who was the *Intendent* of the *Jardin du Roi* since 1739 and one of the individuals most responsible for the aggrandizement of that institution from a minor botanical garden for the study of pharmaceutical plants to a major research institution of the natural sciences. When the transfer of the Réaumur collection took place, Buffon was in the midst of publishing a general natural history. He was halfway through his first section – the quadrupeds – and had not started the second section – the birds. Clearly, it was not to Buffon's interests to have the Réaumur collection mined by someone outside his enterprise. That Brisson was not invited to collaborate with Buffon was a result of the enormous enmity that had existed between Réaumur and Buffon. Brisson was part of a hostile camp and was beyond serious consideration as a collaborator by Buffon. Deprived, then, of the use of Réaumur's collection, Brisson on the advice of the abbé Nollet, relinquished his work in natural history and began the study, and subsequently the teaching of, experimental physics, where he had a successful, if not especially brilliant career. That Brisson never returned to publish in natural history is really quite remarkable if one considers the scope of his ornithology and the fact that in 1760, when it was published, he was merely thirty-seven years old, and had another forty-six years to live. One could attempt to understand this shift of interest in terms of personality and the accidents of professional life. However, equally important was his complete reliance upon a particular collection, of which once deprived it was inconceivalbe to continue working. Had he lived during a later period of history, he might have searched for another collection, however, in 1760 that option was not available.

It is an irony of history that the two major figures of the third quarter of eighteenth-century ornithology, two individuals radically different in personality, in social standing, and in perspective on natural history, should be closely linked by Réaumur's collection.

Georges-Louis Leclerc, comte (as of 1772) de Buffon was the leading natural historian of the second half of the eighteenth century.[15] His writings on ornithology comprise the second section of the *Histoire naturelle, générale et particulière*, which he

Nature lamenting over the Tomb of M. De Buffon, & exhibiting a Portrait of that distinguished Naturalist.

Illustration 3. This undated engraving from the late eighteenth century reflects the enormous popular reputation of Buffon at the time of his death: "Nature lamenting over the Tomb of M. De Buffon & exhibiting a Portrait of that distinguished Naturalist." (author's collection).

began publishing in 1749 and which had run to thirty-six volumes by the time of his death in 1788. This sumptuously printed and illustrated set was, with the exception of a few scientific *mémoires* and a famous *discours* presented at the *Académie Française*, Buffon's *oeuvres complètes,* and therefore was in large part the basis of his reputation as one of the four major philosophes of the French Enlightenment. Daniel Mornet's classic study of private libraries in eighteenth-century France showed that the *Historie naturelle* was the third most popular piece of literature there during the latter part of the century.[16] Buffon's writings were one of the important factors in popularizing science in Enlightenment culture—an achievement that has appeared to historians ignorant of Buffon's scientific importance as a sign of his superficiality.

Like Brisson's *Ornithologie,* Buffon's *Histoire naturelle* was associated with a major collection, the *Cabinet du Roi* at the *Jardin du Roi.* Although it was the largest natural history collection in Europe during the second half of the eighteenth century it was uneven in its holdings and, as most natural history collections of the period, was decidedly weak in birds. Réaumur wrote in 1749 to a correspondent that:

The Cabinet of the Royal Garden is not rich in insects, ores, or birds; the stock of the latter of these consists of sixty or eighty that they had prepared at Strasbourg and which were in large part eaten by moths last year because they do not know how to preserve them.[17]

One can well understand, then, Buffon's desire to secure Réaumur's collection for the *Cabinet du Roi.* Fortunately, he was well placed to effect the transfer of the collection from the *Académie Royale des sciences* to the *Jardin du Roi:* he was the director of the most important natural history institution in Europe; he was treasurer of the *Académie Royale des sciences;* and he had important connections at court. Moreover, the transfer made good sense. The *Académie Royale des sciences* lacked facilities for a major natural history collection, and the *Jardin du Roi* was clearly the center of research in natural history, not the *Académie* which stressed work in the physical sciences.

In the early 1760's, when his plans for the *Histoire naturelle des oiseaux* were being formulated, Buffon in some ways was in a situation similar to Brisson's when he was preparing his ornithology. Buffon had the Réaumur collection and had even drafted Brisson's engraver, Martinet, to provide part of the engravings. Like Brisson, Buffon was profiting from the constant influx of discoveries and observations of travelers, explorers, and correspondents. Buffon, however, had the status and resources to solicit this information on a grand scale; greater even than Réaumur, with whom he had occasionally competed for specimens from mutual correspondents in the colonies. Buffon's efforts had made the *Cabinet du Roi*, which already was the natural repository for gifts to the King from foreign monarchs, into the site for material brought back by government expeditions, bequests to the Crown from collectors, and natural history specimens and observations sent from amateurs, especially in the colonies, who wished to contribute to the "growth of knowledge". Buffon encouraged amateurs by persuading Louis XV to create the honorific title of *Correspondant du Cabinet du Roi*, which in the title-conscious society of the *ancien régime* proved to be a powerful incentive for collecting. Among Buffon's most valuable sources of information were M. Hébert (Dijon), Lefevre Deshayes (Saint-Domingue), Dr. Lottinger (Saarburg), M. Pierre-Augustin Guys (Marseille), Dr. Artur (Cayenne), M. de la Borde (Cayenne), Philippe Commerson (Madagascar), Pierre Poivre (Ile Bourbon), Michel Adanson (Senegal), and M. Baillon (Picardie). Charles-Nicolas Sigisbert Sonnini de Manoncourt (1751–1812) presented Buffon with a large collection of specimens and notes on South American birds, James Bruce (1730–1794) shared with him his knowledge and accounts of his adventures in Ethiopia, and Pierre

Sonnerat (1749–1814) supplied him with a large number of exotics.[18]

Buffon's *Histoire naturelle des oiseaux* could well have been a collection-catalogue natural history, an updated and expanded version of Brisson's ornithology of twelve years earlier. In fact, the original inspiration of Buffon's entire natural history project had been a proposed catalogue of the royal natural history collection. Like Brisson, Buffon was painfully aware that however grand his work might be it still, at best, had to be regarded as preliminary. One man could not possibly acquire the amount of knowledge necessary for a complete picture of the avian world. An expanded catalogue would be a useful step in laying the groundwork for a future ornithology and could be very useful in bringing some order to the chaotic state of nomenclature, a subject which Buffon held to be extremely important. He wrote at the beginning of his article on the bustard:

The first thing that one must undertake when one sets out to elucidate the history of an animal is a rigorous critique of its nomenclature: to specify clearly the different names which have been given to it in all languages and at different times, and to distinguish, as much as possible, the different species to which the same name has been applied. This is the only way that will allow us to make use of part of the knowledge from antiquity and to join it usefully to modern discoveries, consequently it is the sole manner to make true progress in natural history.[19]

Given that Buffon's *Histoire naturelle* had its origin in an initial charge to supply a catalogue for the *Cabinet du Roi;* that he shared the common interests and problems of someone associated with a major collection; and that he too was deeply concerned with the problem of nomenclature; it is all the more striking how fundamentally different his *Histoire naturelle des oiseaux* is from Brisson's *Ornithologie.* Buffon's collections served merely as a starting point for his general encyclopedia of natural history. Indeed, Buffon had nothing but contempt for purely taxonomic

studies which he considered to be sterile exercises, a point he repeatedly made in a number of articles. For example, after describing the odd characters of the newly discovered secretary bird he
stated: "to what class can one relate a being in which are united
characteristics so opposite? Here is another proof that nature,
free in the midst of limits that we think prescribe it, is richer
than our ideas and vaster than our systems".[20] Buffon held that
what was needed in the study of animals was detailed natural
histories of each species. Only *after* such detailed studies were
complete could one begin to construct classifications. He had
pursued just such a course of study with the quadrupeds. Individual articles on the known quadrupeds, published over a period of
almost twenty years, were followed by an attempt to construct a
natural classification based on descent and limited diversification.
With the birds, however, Buffon faced seemingly overwhelming
technical problems. No extant collection was even remotely
adequate for such an undertaking. Buffon estimated that to
provide the minimally adequate description of each species (*not*
including varieties) one would need specimens of a male, a female,
and two juveniles, that is, one would need eight thousand specimens, ten times the size of the collection of the *Cabinet du Roi*
in 1770![21] Even if Buffon possessed such a collection, it would
still be merely a starting point, for in addition to examining the
morphological features of animals, Buffon held that observations
made in the wild were essential to a complete knowledge of
living beings. Such knowledge of the habits of birds was scant at
best, and the prospects of a significant increase were not especially
promising. But a start had to be made, and Buffon attempted to
compile all the available information on birds. Of the existing
literature, he made most use of local faunas. He borrowed quite
openly from Edwards for new species, and often referred to
Brisson, however most often in a condescendingly critical style.
His attempt to clarify nomenclature led him to most earlier

ornithology. However, with the exception of their importance for unscrambling nomenclature, Buffon was generally unenthusiastic about earlier writers, aside from ancient giants like Aristotle. Of the most famous ornithology of the Renaissance, the work of Aldrovandi, Buffon had earlier written: "One would reduce his writing to a tenth of their size if one threw out all that was worthless and foreign to his subject ... often it [Alrovandi's natural history] is mixed with elements of the fabulous and the author is too inclined to credulity."[22] Buffon relied more heavily on his network of correspondents throughout the world, and recently published descriptions of birds by travelers and explorers, such as those resulting from Cook's famous voyages. And, of course, he fully used the specimens he did have. He wrote to his collaborator the abbé Bexon: "Try, Monsieur, to make all your descriptions from the birds themselves, that is essential for precision."[23]

Although Brisson, too, compiled material and paid close attention to his specimens, the scope and intent of his work was quite different from Buffon's. Brisson had set out a thoughtfully articulated artificial system that permitted naturalists to organize the known avian forms. Buffon, in contrast, was aiming for an encyclopedia of knowledge that would be the basis for the discovery of the laws of living beings. This is not to say that Buffon neglected classification. Quite the contrary, for Buffon held that a classification ought to reflect the order of nature. In his work on the birds, Buffon attempted to utilize the insights he had gained while working on the history of the quadrupeds. After describing each quadruped and after considering related factors as geographical distribution, variation, and hybridization, Buffon concluded that closely related species — such as the horse, ass, and zebra — were descendants of an original stock (*premier souche*) which had diversified to a limited extent in time.[24] Although his material was admittedly inadequate to establish a natural classification for birds with certainty, Buffon nonetheless tried to

organize his material as naturally as possible. He grouped birds
into families, described one species in detail and then merely
indicated the distinguishing characters of closely related forms.
The species described in detail was not necessarily to be considered as closest to the primitive stock, but the group as a whole
was intended to be thought of as a related set. The families themselves he grouped, following Ray, by life habits: land birds,
water birds, carnivores, herbivores, etc. It was on this level that the
behavior of the bird was as important as its external characteristics.
In his particular descriptions he occasionally employed behavior
to help distinguish between or among species, however, like
Brisson, Buffon relied heavily on feather color for species distinctions. Great stylist that he was, he nonetheless did not undertake
the thankless task of attempting to depict a complete portrait
of each species and variety. It "would be impossible", he wrote,

unless one used a prodigious number of words, very boring ones at that, for
the description of the colors of birds. There are not even the proper terms
in any language to express the nuances, the hues, the reflections and blend-
ings. Nevertheless the colors are essential characteristics here and often the
only ones by which one can recognize a bird and distinguish it from all
others.[25]

Fortunately, another option was available: color plates. The
famous *Planches enluminées* was begun under the direction of
Edmé-Louis Daubenton (1732–1785) in 1765. When completed
in 1783, it comprised one thousand and eight hand-colored
engravings, of which nine hundred and seventy-three depicted one
thousand, two hundred and thirty-nine birds. At the time it was
a project without equal. René Ronsil stated that "the *Histoire
Naturelle des oiseaux* of Buffon, illustrated by Martinet, is really
the base of ornithological iconography, the point of departure of
a new era for this area".[26] The *Planches enluminées* accompanied
two of the four simultaneously-published editions of the *Histoire*

naturelle des oiseaux; the other two (less expensive) editions had their own uncolored engravings but contained references to the *Planches enluminées*. The quantity and accuracy of the *Planches enluminées* made it possible for Buffon to avoid repetitive analyses of external morphology and feather color. It also turned out to be of particular historical importance since he was writing at a time when collections were in dire threat of almost certain destruction from insect pests, light, damage from alleged preservatives, and sulphur fumigations. A stable collection was not something that could be counted upon, and for that reason the careful illustration of rare specimens and what we now call type-specimens was all the more important for avoiding undue confusion in classification. Practically all of the specimens from the *Cabinet du Roi* have long since fallen into decay.

Buffon's approach to classification differed markedly from Brisson's. At issue, fundamentally, was the tension between a naturalist who aimed at preparing catalogues and keys and one who aimed at uncovering the regularities in nature. The prospectus for the *Histoire naturelle des oiseaux* states the distinction explicitly, even if somewhat polemically:

M. de Buffon scorning to subjugate himself to follow others and not wishing to imitate the puerile pedantry of these nomenclaturists, who give structures from their minds and tables of their petty ideas for the plans of nature, and who make ridiculous associations of beings least made to go together, has delineated a new route, a much simpler plan and one that conforms better to nature's course. In place of dwelling minutely on the details of descriptions, he has studied the nature of the beings he describes, their habits [*moeurs*], their instincts, their practices, their voyages. He has incessantly compared them among themselves and with those most closely related, and it is in thus treating all the parts of natural history that he knew how to extract from them important and useful truths for the physical sciences and for natural philosophy.[27]

Buffon's goal was to reveal design in nature. Unlike the earlier writer abbé Pluche or Buffon's slightly older contemporary

George Edwards, he did not envision design in the context of natural theology, but rather he conceived of it in the context of Enlightenment Deism, such as that espoused by Voltaire. Nature, for Buffon, can be described as a grand tableau filled with interesting and complex relationships. In 1770 he described it in the following terms:

Nature, fully displayed, presents to us an immense tableau in which the categories of beings are each represented by a chain which supports a continual succession of objects sufficiently adjacent and sufficiently similar that their differences are difficult to grasp. This chain is not a simple thread which extends only in length; it is a large woof or rather a bundle which from interval to interval sends out side branches to join with the bundles of another category; and it is especially at the two extremities that these bundles bend, ramify, to reach the others.[28]

In different volumes of the *Histoire naturelle* Buffon employed various metaphors to suggest the overall order and unity of nature. What is of importance in these attempts to express the unity of nature is Buffon's conviction that all phenomena are the result of general laws.[29] His emphasis on nature's uniformity and overall design did not prevent him from seeing nature "as it is" — warts and all. Unlike views of design related to natural theology, which depicted animals in terms of perfect adaptation, Buffon had what we might call an *aesthetic conception* that included *des hombres au tableau*.[30] Although nature is most striking in its harmony and beauty, it has "in the middle of the magnificent spectacle some unheeded productions and some less happy".[31] In his article on the avocette, for example, he wonders how the bird is able to eat with such a monstrous beak, and claims that one must regard such a beak as an extreme form produced by a fecund nature that runs to fullness, hence to occasional extremes.[32] Earlier in his article on the toucans he discussed more fully the nature of "natural monsters" and he wrote that:

One can regard them as monstrosities of the species which differ from individual monstrosities only in that they perpetuate themselves without change . . . . . The true characteristics of nature's errors are disproportion joined to uselessness. All animal parts which are excessive, superabundant, or placed absurdly and which are at the same time more detrimental than useful, should not be placed in the grand scheme of nature's immediate designs but in the small scheme of its caprices, or, if one likes, its mistakes, which nevertheless have a purpose as immediate as the first ones, since these same extraordinary productions indicate to us that all that can be is, and that whatever proportions, regularity, and symmetry reign ordinarily in all nature's works, the disproportions, the excesses, and the defects demonstrate to us that the extent of its power is not at all limited to those ideas of proportion and regularity to which we would like to fit everything.[33]

Buffon's perspective is one that is quite difficult to characterize. It is essentially aesthetic in inspiration. It seeks regularity and order, but recognizes the complexity and enormous variety of nature as well as man's limited means to understand it. The design, imperfectly perceived, is awesome although not necessarily perfect. Buffon's aesthetic conception of nature can, perhaps, be tied to his generally literary approach to natural history. He was one of the great stylists of the century, and his surviving manuscripts, which contain his revisions, reveal the extent of his concern for the literary dimension of his articles. Buffon has been criticized or dismissed because of the highly literary appearance of his articles; as if elegance automatically disqualified one from serious consideration. Yet it was his penchant for *la grande vue* and his aesthetic sensitivity that prompted him to ask fundamental questions about his data. Buffon raised important questions about the relationships of variations, varieties, and species; about their interaction with the environment; about behavior and its relation to the environment and to variation. These considerations, with respect to the quadrupeds and the birds, led him to a dynamic view of nature in which animals had to be understood relative to a constantly changing physical world. Buffon's interest in order went considerably beyond the confines of a "nomenclaturist".

Although his starting point was a museum collection, his ornithology went far beyond the perspective of a curator in asking and attempting to answer basic questions concerning the order in nature.

Although strikingly different, both Brisson and Buffon's ornithology were part of a new approach to the knowledge of birds. The eighteenth century was a period of intense interest in birds. This interest was diffuse, and if it is true that interest in birds permeated eighteenth-century culture, it did so without much of an awareness of birds as a serious object of scientific research. The accuracy, scope, and seriousness of Brisson and Buffon's ornithology mark the beginning in modern times of a scientific study of birds. Brisson's attempts at systematization reflect the existence of the first sizable ornithological collection in Europe. His *Ornithologie* set a new standard for the classification of birds. Buffon's encyclopedia attempted to examine all the available knowledge of birds and to subsume these facts and relations under some general laws of nature. Both works were consciously open-ended and intended to stimulate more studies. From the 1760's on, therefore, ornithology had respectable models and an extensive empirical base. It just would not do to juggle a set of known birds or reported birds as earlier encyclopedists and systematists had done. Descriptions had to be either extensive or accompanied by an accurate illustration and certain basic questions had been posed concerning theoretical issues. What followed in the 1780's was largely a ramification of the start made in the 1760's of a scientific ornithology. The course it took, however, was strongly influenced by a set of technical factors that became important after Buffon finished his *Histoire naturelle des oiseaux*.

CHAPTER III

NEW DATA 1780–1830

Brisson and Buffon each believed that his work, extensive as it was, constituted only a beginning for ornithology. They both held that it would be many years before a complete ornithology would be possible, and they were in agreement that basically what was most needed was empirical information. Buffon had a broader vision as to what would be an adequate empirical base, which included behavior, distribution, and place in the economy of nature, but hoped, along with Brisson, that a solid beginning had been made and would continue to expand. The two could not have been more prophetic. In the fifty years subsequent to the publication of the *Histoire naturelle des oiseaux* (1770–1783) the quantity of information on birds swelled dramatically. In part, this expansion of the empirical base of ornithology was inspired by the impressive start given it in the 1760's and 70's by Brisson and Buffon. The timing of their work was fortunate in that natural history became quite fashionable in the late eighteenth and early nineteenth centuries. Although much of the interest was not as serious or rigorous as later standards would demand, it did create an audience for those few individuals who pursued the study more closely.[1] Of equal or greater importance was the influx of material the collection of which was associated with a new period of colonialization and with the efforts of a few avid collectors who capitalized on the favorable conditions of the time. The major contributions to the growth of empirical knowledge of birds fall into two general categories: regional European and regional exotic species.

European regional faunas modestly extended the number of known European birds, clarified some confusions, added to

27

iconography, and most important, provided and encouraged others to make original observations on the natural history of indigenous and migratory species. In England, for example, the writings of Gilbert White (1720–1793), Thomas Bewick (1753–1828), and George Montagu (1751–1815) illustrate the value of European regional faunas. Gilbert White succinctly stated the importance of such studies when he wrote: "Men that undertake only one district are much more likely to advance natural knowledge than those that grasp at more than they can possibly be acquainted with: every kingdom, every province, should have its own monographer."[2] White's penetrating and sensitive letters to Thomas Pennant and Daines Barrington, which form the core of White's famous natural history, discussed nesting, migration, diet, seasonal changes and other aspects of the natural history of the birds of Selborne. Had he been Buffon's correspondent his observations would probably have been integrated into that savant's general history of birds, and White's name would be no more famous than those of Buffon's other correspondents, like Dr. Lottinger, M. Guys, or M. Hébert. Luckily, White's letters were printed separately, so that in addition to adding some observations to the empirical base, they have served, due to their literary quality and brilliance, as a source of encouragement, incentive, and inspiration to generations of bird watchers, both amateur and professional.

Thomas Bewick's *A History of British Birds* (1797–1804), a work considerably broader in scope than White's collection of letters, was similarly a major force in popularizing the study of birds. A few years before his death, Bewick wrote to his good friend and accomplished bird-watcher in his own right, John Freeman Milward Dovaston (1782–1854), that the praise he received from recognized naturalists was something of a surprise: "No, no, I did not reckon upon their approbation. This is extra — my efforts were directed to the rising generation & my object was to inveigle youth onward by the vignettes to the study of natural history, by which they would be led, in contemplating the works of nature, up to nature's God."[3] Bewick added little to the empirical base of ornithology, but his honest wood engravings made available to a wide public an inexpensive source of iconography for British birds, and encouraged amateurs to partake in field studies. George Montagu's *Ornithological Dictionary; or, Alphabetical Synopsis of British Birds* (1802), and *Supplement* (1813) marks a high point in pre-1830 British ornithology.[4] Although his work did not have the general popularity of White's or Bewick's — nor for that matter the literary or artistic charm of either the aforementioned—Montagu's ornithology established a new standard of accuracy and rigor for British bird books. Most notable was his attention to seasonal, life stage, and sexual variation; exactly those features noted by Brisson and Buffon as necessary for an accurate description. To clear up confusion he even had recourse to the rearing of birds in order to observe the changes in plumage, etc., and in this fashion was able to settle conflicting opinions on issues like whether or not the Hen Harrier and the Ringtail were different species or merely an example of sexual dimorphism. For good reason, then, in 1813 Montagu could write in his *Supplement* that his work was the "most complete history of British Birds extant".[5]

White, Bewick, and Montagu illustrate nicely the contributions of regional studies of European birds. They also reflect the wide range of interests that motivated such studies. Gilbert White's *The Natural History of Selborne* was clearly aimed at documenting a local avifauna as a contribution to the empirical base of ornithology, in terms of local bird lists, migration patterns, and behavior. It also was infused with the enthusiasm of an avid bird-watcher who wished to communicate with those who have equally enjoyed the amusement of field study. Bewick's *History of British Birds* was intended, in a general sense, to be edifying and was in the British tradition of careful natural history done for the sake of natural theology. It was equally a commercial venture — Bewick was primarily an illustrator whose wood engravings were among the most famous of the century.[6] His ornithology, like his *General History of British Quadrupeds*, was primarily artistic and was the source of his income. George Montagu came to his study of birds from his interest in field sport — his first publication was a treatise on gunpowder[7] — and serves as a reminder of the close relation of field sport and natural history.

Motive and accomplishment are, of course, separate issues. However, it is interesting to note how in natural history, especially in the study of local fauna, one sees a nexus of religion, amusement, and commercial interests with local bird lists, iconography, and observations on bird behavior.

In France, the study of local avifauna was somewhat overshadowed by the massive projects of Brisson and Buffon. Additions to the *fauna française* were mostly to be found in the dictionaries of natural history, revised editions of Buffon's *Histoire naturelle*, or other explicit attempts at updating the ornithology of the late eighteenth century. Although there was a large market for books on exotics, there were no serious attempts to produce French or local faunas before 1830, with the exceptions of Philippe Picot de Lapeyrouse's (1744–1818) *Tables méthodiques des mammifères et des oiseaux observés dans le departement de la Haute-Garonne* (1799), and Polydore Roux's (1792–1833) *Ornithologie provençale*, started in 1825 but never completed.

Germany, on the other hand, was the country where regional studies were most impressive in the early nineteenth century. Interest in German ornithology was fostered by Johann Matthaeus Bechstein's (1757–1822) *Gemeinnützige Naturgeschichte Deutschlands nach allen drey Reichen* (1789–1795), in which three of the four volumes deal with birds,[8] and *Ornithologisches Taschenbuch von und für Deutschland oder kurze Beschreibung aller Vögel*

Deutschland für Liebhaber dieses Theils der Naturgeschichte (1802—1812). Although Bechstein did not possess or have access to any major bird collection, nor did he travel much outside his native Thuringia, he did maintain an extensive correspondence and had a good grasp of the literature available. This he supplemented with his own observations of local birds which permitted him to name several new species. Perhaps more important than these were the rustic charm and sensitivity he conveyed in his works. It was this deep feeling for nature that was characteristic of much local fauna, in Germany and elsewhere, in the late eighteenth and early part of the nineteenth centuries. We can see it not only in Bechstein, but in the two other major regional avifaunists of the period, Johann Friedrich Naumann (1780—1857) and Christian Ludwig Brehm (1787—1864).

Johann Friedrich Naumann, who came from a family of amateur ornithologists began publishing in 1820 a "revision" of his father's natural history of German birds, which went far beyond Johann Andreas Naumann's (1744—1826) limited, although excellent, studies. The *Johann Andreas Naumanns Naturgeschichte der Vögel Deutschlands, nach eigen Erfahrungen entworfen . . . aufs Neue herausgegeben von dessen Sohne Johann Friedrich Naumann* (1820—1844) provided a comprehensive account of German birds. Aided by Christian Ludwig Nitzch (1782—1837), who contributed careful anatomical studies, Naumann's twelve volumes helped set a new standard in Germany for natural history.

Christian Ludwig Brehm, at the same time, was publishing ornithological studies of German birds in their environment based upon careful observation. His *Beiträge zur Vögelkunde in vollständigen Beschreibungen mehrerer neu entdeckter, und vieler seltener, oder nicht gehörig beobachteter deutscher Vögel* (1820—1822), served as an inspiration for an entire generation of bird studiers. Brehm was an avid naturalist; he corresponded with and encouraged a wide group of amateurs. He even started a journal to further the study of ornithology: *Ornis, oder der Neueste und Wichtigste der Vögelkunde*, which unfortunately lasted for only three numbers, but which nonetheless foreshadowed later successful publications. Although Brehm's reputation has suffered over the years due to his proliferation of new species names, which caused a great deal of confusion in nomenclature, his prolific and enthusiastic investigation expanded the empirical base of ornithology and attracted others to its ranks.

Outside of the three major regions of ornithological study (Great Britain, France, Germany) one can recognize several other notable works. In Italy, Franco Andrea Bonelli (1784–1830) produced a catalogue of the birds of the Piedmont, Paolo Savi (1788–1871) documented the birds of Tuscany, Fortunato Luigi Naccari (1793–1860) those of the province of Venice, and Girolamo Calvi those of Genoa to name the most well known. Scandinavia had Sven Nilsson (1787–1883), who published the two-volume *Skandinavisk Fauna*, and also had one of the keenest ornithologists of the period, Frederick Faber (1796–1828), who unfortunately died at the young age of thirty-one. In addition to precise observations of Northern birds, Faber asked penetrating questions about distribution, migration, and classification. Friedrich Meisner (1765–1825) and Heinrich Schinz's (1777–1865) work covered the birds of Switzerland which was not surpassed until the work in the later century by Victor Fatio (1838–1906).

As important as regional faunas were, and by the 1820's there were a number of excellent beginnings made toward a detailed knowledge of European birds, it was really in the realm of exotics that one finds the most impressive work in ornithology during the period 1780–1830. In part, this may be a reflection of the empirical base itself: what we now call the Holarctic region, which includes Europe, contains the fewest number of different species of birds and compares very poorly in "oow-aahs" with other regions. The colorful and unusual birds of South America, Africa,

East Asia, and Australia had fascinated Europeans since the earliest explorers carried back parrots and cockatoos. English, Dutch, and French genre painting is an able witness to this fascination. Not surprisingly, natural history collectors prized these show pieces as well as any "new" or "unknown" specimen, *i.e.*, a bird not known to European naturalists. The provincial aspect of considering these birds as "exotic", "unnamed", or "unknown", of course, does not seem to have occured to Europeans, who generally speaking were not characterized at this time by any appreciation of cultural relativity. Brisson and Buffon utilized Parisian collections rich in exotics, and both naturalists were eager to obtain as much information from correspondents and travelers as possible. They had the good fortune of being able to draw on the careful collection and description of men working in some of the ornithologically most interesting areas of the globe: Poivre and Sonnerat in Madagascar and the East Indies, Sonnini in South America, Cook and his naturalists in the Pacific and Australia.

Although hindered by the disruption of revolutions and wars, the arrival of exotics continued in Europe in ever increasing numbers. Traders, explorers, colonials, and *voyageurs-naturalistes* provided a steady flow of new species which broke into a deluge after 1815 when, following the cessation of the Napoleonic wars, nations undertook large scale explorations and surveys. A new period of colonialization, one marked more by maritime, evangelical, and commercial interests than settlements, signaled the magnitude of the commercial expansion of Europe into the far corners of the globe, with France, England, and Holland being the three major colonial powers.[9] The opportunities associated with this expansion, and with the interest it generated appeared limitless.

It was not only the quantity of material sent back that made the influx of exotics so important, but also the quality of materials.

Although a few individuals, such as Poivre, Sonnini, and Sonnerat had been sensitive to the desiderata of ornithologists, in general the material from distant lands available to Buffon or Brisson was a motley collection; haphazardly collected and as much a part of the luxury trade as a part of empirical science. Between 1780 and 1830, however, opportunities presented themselves for knowledgeable or trained individuals to collect birds. The significance of this shift in suppliers of ornithological material was enormous, for the possibility then opened to European naturalists to obtain detailed information on species, sexual and seasonal variation, life histories, distribution, and migration. Not surprisingly, many of these new collectors travelled to the richest bird regions of the globe: parts of South America, Africa, and Australia.

The cost of collection expeditions, which occasionally was substantial, was defrayed in a number of different ways. Some amateurs, such as Coenraad Jacob Temminck's father, Jacob Temminck (1748–1822) or Count Johann Centurius von Hoffmannsegg (1766–1849) financed collectors. Temminck financed François Levaillant's (1753–1824) famous trip to South Africa (1781–1784) from which he brought back two thousand specimens, including many new species. The collection as well as his notes on behavior and habitat served as the basis for his famous *Histoire naturelle des oiseaux d'Afrique* (1796–1808) which, although suspect because of its many errors and occasional flights of imagination, was one of the most important works on African ornithology to appear in the nineteenth century.[10] Von Hoffmannsegg, in addition to corresponding with Francisco Agostinho Gomes (1769–1842) in Bahia, who sent him many Brazilian birds, outfitted and sent Friedrich Wilhelm Sieber (b. 1789) to Pará where he collected in the lower Amazon region for eleven years and contributed towards making the Count's collection one of the most famous of Germany.[11] There were some men of means who desired to explore and collect themselves. William Burchell

(1782–1863), although more famous for his botanical and quadruped finds, brought back nearly three hundred different kinds of birds from his famous expedition in South Africa, the expenses for which were, as he told his readers in the first volume of *Travels in the Interior of Southern Africa*, "entirely and individually the author's".[12] His expedition to South America in the late 1820's similarly was self-financed and highly successful in obtaining bird specimens. Alexander Philip Maximilian Prince of Wied-Neuwied (1782–1867), like many of his contemporaries, took advantage of the favorable political situation in Brazil in the early nineteenth century to explore and collect between 1815 and 1817.[13] The specimens he brought back formed the nucleus of his private museum to which he later added specimens from North America which he collected in the 1830's.

In addition to princes and wealthy gentry a few individuals of more modest means were able to finance collection trips. For example, William Swainson (1789–1855) collected in Brazil from 1816 to 1819 by supplementing his military pension from various sources and by obtaining letters of introduction that gave him access to official colonial hospitality and cooperation. His life is an interesting example of the formidable difficulties that confronted anyone trying to make a living as a naturalist in the first half of the nineteenth century. From his autobiography[14] and his correspondence[15] one can glean the story of Swainson's ever necessary preoccupation with staying solvent by writing, drawing, and collecting ventures, none successful enough to provide a stable and secure existence. Nor could he obtain one of the very few posts available to naturalists, such as the Keeper of Zoology at the British Museum for which he unsuccessfully campaigned. Little wonder that he eventually emigrated to New Zealand.

Swainson, more than anything else, desired some governmental or institutional base for his work; not only for the financial

security but also for the legitimization that it implied. In his
autobiographical essay, he wrote of his failure to publish an
account of his South American travels that: "I was discouraged
by the idea, that the unpatronised researches of an unknown
individual might probably be thought insignificant, when compared
to those of naturalists sent out by governments."[16] Fortunately,
not all young aspiring naturalists shared Swainson's fate. Governments and public institutions sponsored a few naturalists during
the early nineteenth century. The support was minimal compared
to the grand explorations of Cook, Bauhin, and the circumnavigations of the 1820's, but in "cost-effectiveness", to use a word, but
not a concept, wholly unknown at the time, the money was well
spent. The individuals were carefully chosen and brought back
myriads of specimens; both collected and purchased. For this
reason all public institutions that could aspire to government
patronage desired to have their own qualified collectors in the
field. The most important and best establishment for such collecting was attained by the professors at the *Muséum* in Paris. On
February 3, 1819, at a meeting of the staff of the *Muséum,* the
Secrétaire général of the *Ministre de l'intérieur* informed the
professors of the *Muséum* that the budget contained twenty
thousand francs to create "a school for young naturalists destined to make voyages to divers parts of the world".[17] The sum
soon grew to twenty-five thousand francs per year and was used to
train, equip, and cover the expenses of collecting for about ten
voyageurs-naturalistes. Considering that the (admittedly underpaid)
professors of the *Muséum* each received an annual salary of five
thousand francs, the grant was generous. It was not munificent,
for an equal sum would have been necessary to maintain a fashionable, although not extravagant, bachelor living in Paris for a year.
There was no shortage of applicants. In the first year there were
thirty-two complete dossiers, which included comments from the
interviews given by the *Muséum* staff. Part of the grant was also

used to help defray the costs of other more established naturalists and to cover the costs of purchasing specimens. The program and subsequent activity of its students was so successful that the *Muséum* soon had little need to train new people and the annual grant became a general fund for *voyageurs-naturalistes* and purchase. With the aid of government support, the *Muséum* was able to send men into the field in areas where it wished to strengthen its collection and where political circumstances permitted. Missions were sent to the Cape of Good Hope, South America, Australia, West Africa, Madagascar, North America, India, etc.[18]

The French funding of naturalists was not unique, and the value to a museum of having its own collectors was obvious. A number of governments contributed to support the quest for exotic specimens which were popular not only among ornithologists but also among the general public whose imagination was stimulated by numerous popular travel accounts to exotic places. The Imperial Court Cabinet in Vienna was richly endowed and financed a few naturalists, most notably Johann Natterer (1787–1843) who collected in Brazil for seventeen years and assembled a collection of twelve thousand specimens. Germany had no central museum, although the new University of Berlin, with its museum (1810), in many ways acted as the model. The Prussian government helped to enrich the Museum's collection by financing such extensive expeditions as those of Friedrich Sellow (1789–1831) in South America, and Friedrich Wilhelm Hemprich (1796–1824), and Christian Gottfried Ehrenberg (1795–1876) in the Middle East and Northeast Africa.[19] The Dutch government which had extensive interests in the East Indies exploited the natural history of its exotic possessions. Between 1815 and 1822 Carl Reinwardt (1773–1854), as "Directeur tot de zaken van den Landbouw, Kunsten en Wetenschappen op Java en de naburige eilanden" [Director of matters pertaining to agriculture, arts, and sciences on Java and the neighboring islands] collected many new species which were presented to the *Rijksmuseum van Natuurlijke Historie* (1820).[20] Much more important, the Dutch government also financed a group of eminently qualified collectors known as the *Natuurkundige Commissie* to explore the same region. The *Commissie* included Heinrich Kuhl (1797–1821), one of the most promising ornithologists of the period, and Johan Coenraad van Hasselt (1797–1823), both of whom died shortly after arriving (1820) in the East Indies, but *not* before they could send back from Java two thousand bird specimens! They were followed in 1825 by Heinrich Boie (1794–1827) who, like Kuhl, died shortly, Heinrich Christian Macklot (1799–1832), Salomon Müller (b. 1804) and an artist, Pieter van Oort († 1834). Other naturalists followed until 1850 when the Commission was disbanded. Of the four listed who went out in 1825, only Müller returned alive (in 1837) — with a collection of six thousand, five hundred bird skins and a large number of skeletons, nests, and eggs.[21] The Portuguese Crown financed expeditions in the late eighteenth century to its overseas territories such as Brazil, Angola, and Mozambique. Much of the collections that were made, however, were taken to Paris by Etienne Geoffroy Saint-Hilaire during the French occupation of Lisbon in 1808.[22]

Museums also purchased specimens when possible. The sale of natural history specimens had been a small but lucrative trade for over a century, but generally it was more geared to the small amateur *cabinet d'histoire naturelle* or luxury trade than any serious scientific endeavor. With the growth, as will be discussed in the next chapters, of major collections — public and private — and the increased commercial contacts throughout the world, especially after the Napoleonic wars, the opportunity for a more "serious" commercial naturalist arose. The two most famous ones were Leadbeater and Verreaux. In London, starting at the turn of the century, Benjamin Leadbeater had a natural history and taxidermy agency near the British Museum and supplied and/or prepared specimens to that institution as well as to many of the major collections in Great Britain and the rest of the world. Across the channel the Maison Verreaux provided similar services, but on a larger scale. The agency was founded by Pierre-Jacques Verreaux in 1800. His three sons, Jules, Edouard, and Alexis, later transformed the modest establishment into the foremost one of its kind. They made extensive collecting trips principally in the Cape Colony, where they opened a branch which sold to collectors who had come there (often on a short stop) as well as to collectors in Europe. Jules also later in the 1840's traveled to Australia and Tasmania to collect for the *Muséum*.[23]

The numerous naturalists and trained collectors who scoured the unknown portions of the globe for profit, glory, or curiosity in search of new species or who patiently observed the habits, distribution, and variation of exotic birds, made significant contributions to the empirical base of ornithology. Governments aided this activity by direct or indirect patronage and by diplomatic liaison. Some diplomatic posts were staffed by naturalists, for example, Georg Heinrich von Langsdorff (1774–1852), the Russian consul to Rio de Janeiro, who actively collected in Brazil and who served as an important contact for other collectors. Governments also on occasion mounted large scale expeditions, which could accommodate naturalists or individuals who served as naturalists, and the result was often a major collection. The most important of these "windfalls" to ornithology were the collections acquired by the famous French circumnavigations of the globe during the late eighteenth and early nineteenth centuries. Cook's voyages of discovery had stimulated the French in the late eighteenth century. As a consequence, Jean François de Galaup, Comte de LaPérouse (1741–1788) was sent in the *Astrolabe* and *Boussole* (1785–1788) with a sizable scientific contigent to make observations. The expedition was not wholly scientific for LaPérouse was charged to look after certain economic and political interests.[24]

After the mysterious disappearance of LaPérouse's ships in 1788, Bruny d'Entrecasteaux (1739-1793) was sent by the National Assembly in the *Recherche* and the *Espérance* (1791—1794) to search for his missing countrymen. His expedition was carefully equipped and prepared to make observations and collections, but was also ill-fated. He died at sea and his ships were later interned in Java where his collections were seized and a sizable portion of his men died of disease.[25] Nicolas Baudin (1754—1804), who led the only other major French exploration until the end of the Empire, had better luck on his voyage to Australia in the *Géographie* and the *Naturaliste* (1800—1804) which were well staffed with scientists, artists, and instruments, and netted one of the largest natural history collections to date.[26] The nine hundred and twelve bird specimens represented two hundred and eighty-nine species, one hundred and forty-four of which were new.[27] Many of these were later described by Vieillot in the second edition of the *Nouveau dictionnaire d'histoire naturelle*.

Baudin's expedition, although highly successful from a scientific point of view, was torn by internal strife. The bitter disputes between the scientific staff and the naval personnel led to the decision that henceforth the collection and observation of natural history on navy ships would be entrusted to medical officers.[28] This change in policy not only reflected an internal ruling occasioned by a very unfortunate circumstance, but also a growing professionalization of the French navy as well as a shift in emphasis in the purpose of major naval expeditions. The great voyages of the post-Napoleonic period were increasingly undertaken with an eye towards political and commercial value. "Exploration for its own sake, for science, for the advancement of geographical knowledge and of navigation, played a continually decreasing role."[29] Expansionist policies, closely tied to mercantile interests, can be detected in most of the European expeditions. Nevertheless, the French great voyages brought back very sizable collections due to the diligence of several dedicated medical officers. The scientific dimension of the voyages, however, was clearly secondary. René Lesson (1794–1849) and Prosper Garnot (1794–1838) even felt it necessary to tell their readers in the zoological portion of the report on the expedition of the *Coquille*: "We must again add that although our collections were large and varied, they were the result of our own individual resources and that they occasioned no expense to the expedition."[30] To be sure, such dedication was acutely appreciated by the scientific community. In a report made to the *Académie des sciences*, Arago stated:

The *Muséum* of the *Jardin du Roi* has not only been enriched by the efforts of MM. Quoy and Gaimard, surgeons of the expedition, with a large number of very rare objects which until now have been lacking in our collections, but they have also obtained a considerable number of entirely new species. The zeal of these two voyagers merits, moreover, the highest praise; for they are not naturalists by profession and could only bring to their research that general education which includes zoology in general. They themselves have prepared, with an indefatigable ardor, the animals they collected, and co-jointly with M. Gaudichaud, pharmacist of the *Uranie*, they have offered, with a noble disinterest, to the *Muséum* a number of curiosities which they acquired during the voyage.[31]

In terms of ornithological data, the most important French voyages in the 1820's were Louis de Freycinet's in the *Uranie* and the *Physicienne* (1817–1820) [material collected by Quoy and Gaimard]; Duperrey's in the *Coquille* (1822–1825) [material collected by Lesson and Garnot]; Hyacinthe Bougainville's in the *Thétis* and *Espérance* (1824–1826) [material collected by Busseuil, described by Lesson]; and J. Dumont d'Urville's in the *Astrolabe* (1826–1829) [material collected by Quoy and Gaimard].

The major voyages of the 1820's were French, however the British were not idle, although their explorations of the 1830's are better known. Of ornithological significance was F. W. Beechey's in the *Blossom* (1825–1828) and the land expeditions of Henry Salt to Ethiopia (1809–1810) and John Franklin's two expeditions

(1819–1822, 1825–1827) to the Far North of the New World accompanied by John Richardson who collected material.

In the eighteenth century, the major portion of exotic material was acquired by European naturalists from travelers to or residents in colonies. The trained collectors and the naturalists on large-scale expeditions during the late eighteenth and early nineteenth centuries soon surpassed these colonial suppliers. However, the colonies continued to contribute to the increase of ornithological information, and on a large scale. Excellent work was done by residents of colonies (or ex-colonies in the case of the United States) and sometimes was more on the order of the regional fauna work done in Europe than the mere collection of exotic material. By the late eighteenth and early nineteenth centuries some of the colonies had been settled for over a century and had been the object of study all during that time.[32] It should not be surprising, then, to note that colonial naturalists, in addition to supplying specimens, were beginning to make original contributions. In North America, for example, the works of Alexander Wilson (1766–1813) and John James Audubon (1785–1851) contributed to the knowledge and iconography of that continent's avifauna. Wilson's nine-volume *American Ornithology* (1808–1814), in addition to describing thirty-nine new species and differentiating twenty-three others from European species with which they had been confused, is a rich source of original natural history observations on American birds. Audubon's magnificent *The Birds of America* (1826–1838), as an artistic work, ranks with the splendid bird art books of France, England, and Germany.

Individuals like Wilson or Audubon, although important, were unusual. A greater source of information on birds came from amateurs associated with the great trading companies and or military establishments in the colonies, especially in the British and Dutch empires. The East India Company in 1801 founded a museum in London, the zoological portion of which consisted

... of specimens in all departments of the Science, from the Company's Oriental possessions, contributed by public servants who have been attached as Naturalists to Missions and Deputations on behalf of the Indian Government, or by gentlemen of the civil and military services as presents to the Honorable Court of Directors.[33]

Thomas Horsfield's catalogue of the ornithology collection done in the middle of the century, just before the company was disbanded, documented the contributors to the Museum and shows collections from Ceylon, India, Sumatra, Java, Siam, what we now call Indochina, the Himalayas, etc. Some of the collections were sizable. For example, the Marquis Wellesley (1760–1842) presented the Company with two thousand, six hundred and forty folios of natural history paintings. He also encouraged others to collect. When he was Governor General of Fort William (1798–1804), he issued the following minute:

The knowledge hitherto obtained in Europe respecting certain branches of the natural history of the continent of India and of the Indian isles is defective. Notwithstanding the progress which has been made within the last twenty years in the prosecution of scientific enquiries connected with the manners, produce, and antiquities of this part of Asia, many of the most common quadrupeds and birds of this country are either altogether unknown to the naturalists of Europe, or have been imperfectly and inaccurately described.

The illustration and improvement of that important branch of the natural history of India, which embraces an object so extensive as the description of the principal parts of the animal kingdom, is worthy of the munificence and liberality of the English East India Company, and must necessarily prove an acceptable service to the world.

To facilitate and promote all enquiries which may be calculated to enlarge the boundaries of general science, is a duty imposed on the British Government in India by its present exalted situation, and the discharge of that duty is in a more especial manner required from us, when any material addition can be made to the public stock of useful knowledge without involving considerable expense.

The Governor-General entertains a confident persuasion that, with the facilities which we now possess for the collection of accurate information from every part of India, the natural history of this quarter of the globe may be greatly improved and extended within a comparatively short period of time, without involving the necessity of any material charge on the public resources; but this desirable object will never be attained, unless it shall be made the duty of some public officer, properly qualified for this service, to collect information, and to digest and publish the result of his researches. Under these considerations the Governor-General has had it in contemplation, for some time past, to select a person conversant in natural history to be employed in the cultivation of that useful science, in the Asiatic possessions of Great Britain.

The knowledge, the learning, and the former habits of Dr. Francis Buchanan, have rendered him perfectly competent to the performance of this task, and the Governor-General accordingly proposes, that Dr. Buchanan be directed to collect materials for a correct account of all the most remarkable quadrupeds and birds in the provinces subject to the British Government in India, and to extend his enquiries as circumstances shall admit, to the other divisions of this great continent and to the adjacent isles.

To facilitate the discharge of this duty, the Governor-General has provided an establishment at Barrackpore, where the quadrupeds and birds which may be collected for Dr. Buchanan will be kept until they shall have been described and drawn with that degree of attention to minute distinctions, which is essentially necessary for the purposes of the natural historian.

The Governor-General proposes that circular orders be transmitted to the principal civil and military officers at every fixed station under this Presidency, requiring them to instruct the medical gentlemen under their authority to correspond with Dr. Buchanan on this subject, and to reply with dispatch and accuracy to Dr. Buchanan's letters; that the principal civil and military officers be further directed to authorize their medical officers to solicit assistance and information from all officers of government under their authority, whether European or native, and that they be required to instruct all persons employed in the service of Government to afford to their medical officers the necessary assistance in procuring such animals as may be required, to communicate the most accurate information which can be obtained from the most intelligent persons in the vicinity respecting their natural history, and to furnish such aid as may be necessary for the conveyance of the animals to the Presidency.

The Governor-General proposes that the Right Honourable the Governor in Council of Fort St. George, the Honourable the Governor in Council of Bombay, and his Excellency the Governor of the British possessions on the Island of Ceylon, and the Lieutenant-Governor of Prince of Wales's Island, be requested to direct the proper officers under those governments respectively to correspond with Dr. Buchanan and to afford every practicable assistance to Dr. Buchanan in the prosecution of his researches within the limits of their local authority, and that similar orders be sent to Malacca and to Bencoolen . . . .

The Governor-General proposes that the observations of Dr. Buchanan on such subjects of the natural history of animals as may be collected, together with the drawings of each subject, be transmitted once in each season to the honourable the Court of Directors, with a request to the honourable Court to direct the publication of the work, in such manner as they may deem most proper.[34]

Although the results of this particular minute were less impressive than the enthusiasm that motivated it, it does reflect the attitude of a number of important British colonials, often not fully appreciated by naturalists in London.[35] Some of the East India Company personnel such as Major General Thomas Hardwicke (1755–1835) contributed collections of specimens and illustrations to the East India Company Museum and to other institutions: the Linnean Society, the Zoological Society, the British Museum, and the Museum of the Asiatic Society at Calcutta. Among the most valuable possessions that Hardwicke brought back were drawings of animals. This collection, as those of others from Asia, was quite important in as much as often the specimens brought from that portion of the world were poorly preserved and soon deteriorated. A number of types are based on Hardwicke's drawings.

Although the British, Dutch, and French were the leading colonial powers in the late eighteenth and early nineteenth centuries, they did not have a monopoly on collections. The old Spanish and Portuguese empires, which were increasingly moribund, were primarily a source for international collectors when they were permitted access, however, all work was not done by foreigners. There was, for example, Félix de Azara (1746–1821) a Spanish military engineer, who spent twenty years in South America, and who published upon his return *Apuntamientos para la historia natural de las Paxaros del Paraguary y Rio de la Plata* (1802–1805) which contained a wealth of original observations

on natural history and was utilized all through the nineteenth
century (mostly in the French translation of 1809 by Sonnini).[36]
International religious orders, such as the Jesuits[37] also collected
and published faunas in South America, as well as in other distant
lands. The most famous was that of the Italian Jesuit Giovanni
Iganzio Molina (1740–1829) who lived for years in Chile and
ultimately published the important *Saggio sulla Storia naturale
del Chili* (1782).

The period from 1780 to 1830 was clearly a period during
which the empirical base of ornithology increased dramatically
both in quantity and quality. Europeans were producing local and
regional faunas which in an age before specialized journals were
the main loci for the description of new species or the classification of previously confused descriptions or nomenclature. These
early faunas also laid the foundation and supplied much of the
iconography for the extensive regional works done later in the
century. Into what were considered the more remote regions of
the globe, major collectors financed expeditions to bring back
exotics, and some wealthy and/or enterprising naturalists embarked on their own explorations. The French and Dutch governments
financed sizable enterprises, devoted to procuring natural history,
as did, to a lesser extent, the courts in the German-speaking parts
of Europe. The British relied more on its amateurs through the
Empire to enrich its collections. Occasional windfalls, like those
from the French circumnavigations brought in even more.

What sort of general statements can one make about this new
empirical base? For the most part it consisted of bird skins. Major
collectors at this period still attempted to have their specimens
mounted, but the number of specimens involved was overwhelming and institutions like the *Muséum* fell hopelessly behind in
mounting skins. Similarly, private collectors, like Burchell, often
had cases that were sealed and left unexamined.[38] In fact, most of
the material that was collected, especially in the 1820's (*e.g.,* the

French voyages) was not described for well over a decade. In addition to skins, drawings or descriptions were, in many cases, the form the data took. This was especially true for the material coming from India and the Orient, or from naturalists like Azara who did not or could not return with a major collection of specimens.

But there is a lot more to ornithology than external morphology. An effort was made, especially by French naturalists, to obtain skeletons when possible. The *Muséum's* commitment to comparative anatomy led them to build a major collection.[39] Similarly, John Hunter (1728–1793) built a fine collection of bird skeletons which he displayed in his museum and which ultimately passed to the Royal College of Surgeons of London.[40] By contrast, there was among the data a paucity of information on distribution, migration, and behavior. Most of what existed was found in the European regional faunas, especially the brilliant work of people like Faber and White. Associated with the field observations were occasionally collections or descriptions of eggs and nests. Although some collectors of exotic material were as interested in the natural history of birds as the authors of European local faunas, their work is lost among a sea of thousands of specimens sent back in cases and containing a minimum of information.

If one reads widely among the printed documents from the period 1780 to 1830 one can find, in addition, a number of other observations on birds: extensive discussion on the breeding of domestic fowl, delightful observations on parlor birds, some extensive comparative anatomy and some rudimentary physiology. But what one must keep in mind is that these "other" data were not thought to belong to the realm of ornithology, and indeed their focus was elsewhere: practical agronomic questions, the laws of general physiology, the laws of comparative anatomy, etc.[41] The definition given to ornithology by Buffon and Brisson had "taken". This will be more evident when we discuss the nature of the ornithological works produced in this period. Before doing

that, however, it will be necessary to see how what was recognized
as ornithological data was housed, *i.e.*, the development of ornithological collections.

CHAPTER IV

LOCI OF NEW DATA: COLLECTIONS 1786–1830

The half century that witnessed the enormous influx of ornithological data also witnessed a corresponding transformation in natural history collections. Until the end of the 1790's these collections continued as they had been for nearly a century, that is, in the amateur *cabinet d'histoire naturelle* tradition. They were general collections of natural history objects: shells, insects, minerals, a few quadrupeds and some birds, and often formed but a subsection of a larger general collection of *objets d'art,* antiquities, and books.[1] The owner was normally not a naturalist, nor did he or she publish anything other than an occasional catalogue, usually a sale catalogue.[2] The conception of a natural history cabinet therefore was more that of a collector's than of a savant's, and consequently aesthetic considerations were as important as scientific ones.[3] These aesthetic concerns are reflected in the enormous attention paid to display. The third edition of d'Argenville's popular treatise on shells and an important source of information on contemporary natural history collection suggested, for example, the following:

Those who have a considerable number of birds can display them in an enchanting manner by placing them on the branches of an artificial tree, which has been painted green and placed at the back of a grotto-like niche, with a small fountain, in which the water, instead of from a spring comes from a pump or from a small lead cistern placed at roof level to receive rainwater.[4]

The design was not fanciful. D'Argenville himself had a celebrated cabinet which Peter Simon Pallas (1741–1811) described in 1763 in a letter to the avid collector Emmanual Mendez da

Costa (1737–1787): "D'Argenville has some great trees in his Museum, with leaves made of Iron plates, and their branches full of stuffed up birds. The reeds covering the roots of those trees are full of Toads, Lizards . . . ."[5]

Pallas continues in his letter to make a general observation on Parisian collections:

Collecting natural Curiosities is now in vogue to that degree that no body is thought du bon ton, who has no collection. The decorations of some Cabinets are more expensive than the Curiosities themselves, and taste so much of that gout manqué (or outré) . . . that the Collections seem more like Rary-shows, than anything of a Scientific nature.[6]

Natural history cabinets, in fact, were considered to be part of the sights of a city and were important enough to be included as part of a "grand tour". Thiéry's *Guide des amateurs et des étranges voyageurs a Paris* (1787), one of the more well-known guides to Paris, lists forty-five *cabinets d'histoire naturelle* of note.

Access to private natural history cabinets was usually by introduction or personal acquaintance. A few museums were open to the public for a fee. These were either commercial enterprises or were owned by individuals whose passion for collecting surpassed their means. Among the most well known in this latter category, and for a time quite fashionable, was Sir Ashton Lever's (1729–1788) museum in which scattered among other curiosities were many thousands of bird specimens, many quite rare. Typical of the museums of the period it emphasized the exotic, and while not quite down to the level of "Rary-shows", it did not cater to an audience of serious naturalists.[7] Madam D'Arblay had the following account in her diary:

Tuesday, December 31st [1782]. I went this morning with my dear father to Sir John Ashton Lever's, where we could not but help but be entertained. Sir Ashton came and talked to us a good deal. He may be an admirable natural*ist*, but I think if in other matters you leave the *ist* out you will not much wrong him. He looks full sixty years old, yet he had dressed not only two young men, but himself, in a green jacket, a round hat, with green feathers, a bundle of arrows under one arm, and a bow in the other, and thus accoutred as a forester he pranced about, while the young fools, who were in the same garb, kept running to and fro in the garden, carefully contriving to shoot at some mark, just as any of the company appeared at any of the windows.[8]

In contrast to Lever, was the early nineteenth-century proprietary museum of William Bullock in London which in addition to entertaining the public made a serious effort to instruct as well. Unlike the "wild beast show" at Exeter Change in the Strand or the eccentric Leverian collection, Bullock carefully labeled his specimens and displayed them in a scientific manner. William Jerdan, in his *Men I have Known* (1866) recalled the opening of Bullock's Museum:

Up to that date there was nothing of the sort . . . . The British Museum was not, in those days, a place of popular resort. The Leverian Museum, in the Blackfriars Bridge Road, was a most heterogeneous medley of stuffed animals, without order or classification, and savage costumes, weapons, and products from the Pacific Ocean, or elsewhere in Asia, Africa, or America, as such curiosities were picked up by adventurous navigators and exploring travellers. In a visit to it a few desultory facts might be gathered; but as a means for solid or lasting instruction, its miscellaneous and aimless character rendered it useless. Mr. Bullock's collection was quite the reverse of this — admirably preserved and scientifically arranged. After three or four experimental years in its original locality, it was transported to the Egyptian Hall, then finished for its reception, and not fewer than 32,000 subjects of animated nature were skillfully grouped and conveniently displayed within its walls. The town was absolutely astonished by the individual acquisition of so vast and marvellous a treasure and crowds soon availed themselves of the privilege of reading its lessons . . . . In one department were seen the quadrupeds, as natural as life, and as they would appear in a real Indian forest, with its rocks, caverns, trees, and all other adjuncts congenial to their habits and habitats. In another direction, 3,000 birds were set up with similar accuracy, and attended by well-selected acessories, so as to afford sufficient ideas of their motion, food, and mode of feeding, and peculiarities of every description.[9]

Bullock's museum was quite different, then, from the general pattern of natural history collections before 1800 and reflects the contrast well. This is not, of course, to say there were no serious collections before 1800. Marmaduke Tunstull (1743– 1790) assembled a major collection of birds which he used in the 1770's for his list of British birds, *Ornithologia Britanica* (1771).[10] Similarly, the Naumann collection as well as those of Pallas and Latham were serious and intended for scientific use. They constituted, however, but a small portion of the natural history collections of the period.

In the first three decades of the nineteenth century that small portion grew dramatically and soon eclipsed the rest. The most prominent for much of that time and the model for many was the collection of the *Muséum National d'Histoire naturelle,* the new name given the *Jardin du Roi* after its reorganization in 1793. Buffon had tried to establish there the largest extant collection of birds. His successors strove for the same goal and with spectacular results.[11] It was the main repository for collections brought back by government expeditions, for gifts, and for bequests to the nation. Also, the professors of the *Muséum* continued Buffon's method of encouraging foreigners and colonials to send material by designating them as *correspondants.*[12] During the revolution emigré collections were transferred to the *Muséum* for incorporation or dispersal among provincial public institutions, and a number of important collections were taken by the *commissaires pour la recherche des objects de science et d'art dans les pays conquis* as war prizes, including the splendid cabinet of Stadtholder,[13] which contained many new or rare exotics from the Dutch colonies.[14] The *Muséum* enjoyed enormous official support, for it drew on the rising public taste for natural history and promoted that interest by lecture series and by its popular menagerie. The government's annual grant to provide for *voyageurs-naturalistes* is but one sign of the support the *Muséum* enjoyed.

Not only did the French national natural history museum continue to expand its bird collection, but it also reorganized its holdings. The collection assembled under Buffon's administration had fallen into decay. The most spectacular had been mounted and placed on display, but the majority were packed in cases where they were prey to insect pests. Under the able administration of Etienne Geoffroy Saint Hilaire (1772–1844) the collection was renovated and expanded. By 1809 he could report that the bird collection had grown from four hundred and sixty-three (in 1793) to three thousand, four hundred and eleven specimens.[15] The collection included those specimens brought back from Australia by Baudin, the birds collected by Geoffroy on his expedition to Egypt with Napoleon, a notable collection of exotics that Geoffroy had acquired in Lisbon during his mission to that country after its conquest, the Statholder's collection, as well as over six hundred specimens from correspondents.

Not only did the *Muséum* continue to receive new specimens, but it also had the staff to prepare these specimens for display.[16] Louis Dufresne (1752–1832)[17] who began his career as a naturalist on the *Astrolabe* (fortunately he returned early, otherwise his career would have terminated in that ill-fated voyage), was appointed *aide-naturaliste* in 1793. He was among the most famous taxidermists of the century, partly because of his tasteful displays but also in part due to his publication on the subject. The great peril for bird collections in the eighteenth century was insect pests, as Réaumur had clearly indicated in the middle of the century. Although he had experimented with a large number of substances, Réaumur had not succeeded in his search for an effecitve preservative and had relied on periodic sulphur fumigations.[18] Buffon, too, relied on sulphur, and although the method succeeded in saving the mounted birds from insect attack it did so at the cost of ruining the specimens. It was due to this damage that Geoffroy had to replace most of the collection in the late eighteenth century. Merely a handful of specimens from the "Buffon collection" exists today. The problem of preserving birds from insect attack was solved[19] by Jean-Baptiste Bécoeur (1718–1777) an apothecary in Metz and owner of a magnificent bird collection. Bécoeur developed an arsenical soap that protected skins without destroying them. Although Bécoeur kept his recipe a secret during his lifetime, in the hope of benefiting financially from it, it somehow passed to the *Muséum* and became the accepted method of preparation. Dufresne popularized the arsenic soap in an article on taxidermy that he wrote for the *Nouveau dictionnaire d'histoire naturelle* (1803–1804) and then in a separately published extract (1820) from the expanded second edition of the *Nouveau dictionnaire* (1816–1819). The application of an arsenical soap to well-cleaned skins provided effective protection from insect pests. Dufresne also carefully labeled each specimen. Unfortunately, like all of his contemporaries, he also had them mounted, and over the years air, light, heat and dust have taken their toll.

The *Muséum* did not have a resident ornithologist, however its collection was well arranged and generally accessible. French naturalists as well as foreign ones utilized it in describing new species, preparing monographs, bird lists, etc. It was this accessibility coupled with the scope of the collection that made it the metropole for ornithology, and the standard by which other museums were judged.[20]

Not only was the bird collection of the *Muséum* accessible and enormous by contemporary standards, it contained over six thousand specimens in good condition representing over two thousand, three hundred species,[21] but it also was a permanent collection. In this sense it contrasted strikingly with the majority of the collections of the eighteenth century which were in private hands and which posed the critical problems of disposal, and usually dispersal. Take the Lever collection. It was not a financial success, and after a number of appeals for public support or for private patronage failed, Lever offered it for sale to the government for the nation. The British Museum, the obvious repository for such a collection, however, was already overcrowded, and there was little enthusiasm in government for such expenditures. Da Costa rather caustically noted in a letter that

In regard . . . the Encouragement given to Sir Ashton Lever for his laudable undertaking which reflects on honour to England, my meaning [?] was that The English Nobility and Gentry seldom encourage either English Worth or Science but their heads and hearts lavishly encourage and reward all foreign (or french if you please to read it so) Whores and Pimps numbers of Whom are more encouraged in one day than Sir Ashton or any Scientific or Beneficial Englishman would experience in his whole life.[22]

Lever decided to sell his entire collection by lottery.[23] The winner, Mr. James Parkinson, a dentist who had bought the winning ticket for one guinea (one of the eight thousand sold of the thirty-six thousand issued of which the remaining twenty-eight thousand were in Lever's possession!), also was not able to make the museum show a profit, and sold it in 1806. The sale[24] took over two months and contained seven thousand, eight hundred and seventy-nine lots. Over two hundred birds were purchased for the Imperial collection in Vienna.[25] Bullock, Swainson, Lord Stanley, and Latham were among the major bidders. And so the famous Lever collection, which among other things contained much of the Cook material, was dispersed. Thirteen years later Bullock sold his museum by public auction.[26] Representatives from all the major public and private collections were at the month-long sale and material went to Leyden, Paris, Vienna, Berlin, Edinburgh, as well as to dealers and small collectors.

Although European naturalists were quick to seize the opportunity of obtaining rare birds at public auction, they were equally disturbed by seeing major collections dispersed. This was so primarily because of the scientific value of these collections and the value of having type-specimens in a known, fixed location. For anyone writing a regional fauna, a bird list, a monograph, or a general treatise large collections were essential. Unless one had the means to travel extensively there was an enormous practical value in having an accessible and permanent collection in some known and not too distant location. We have seen what happened to Brisson when deprived of the Réaumur collection. The periodic breakup of major collections made it difficult, if not impossible, for naturalists, without the means of assembling a large collection themselves, to consider any extensive ornithological work and made it equally vexing to follow up or check on a work based on a specific collection. Imagine how bleak this quotation from Bullock's sale catalogue would have appeared to a young aspiring continental ornithologist who had traveled to London to consult Bullock's museum:

The almost exclusive command of the seas, during a protracted war, successively filled this country from every part of the world with the most novel and extraordinary specimens in this branch of Natural History which generally centered in this Museum, and formed an important part of its extensive attraction. There are many thousands of birds unknown (chiefly owing to what we have already said of the maritime nature of the war) to continental Naturalists, and for which names are not to be found in the Linnaean classification.[27]

Perhaps even more important was the issue of type-specimens.[28] The beginning of detailed regional studies and the influx of exotic material focused attention on new species, and consequently the importance of an accurate identification of the particular specimen used by an author to name a new species was soon perceived. Type-specimens came to be carefully labeled and became a valuable part of collections because they could be consulted as reference material for re-examination at a later date. The breakup of collections seriously impeded the process of checking type-specimens. John Latham (1740–1837), who in his *A General Synopsis of Birds* (1781–1785) had carefully indicated a collection for most of his descriptions, in his *A General History of Birds* (1821) had to sadly preface his work with the following comment:

It will be observed in many cases, that birds are said to be in various cabinets, not now in existence — but it must be remembered, that at the time of the first penning the Synopsis, and long after, the Leverian Museum was in full preservation. Many subjects also, referred to in the British Museum, have since fallen into decay; and the very numerous and choice articles then in Mr. Bullock's noble collection are now dispersed. The reader has, therefore, to rely on the author only for the description.[29]

The *Muséum* exemplified the value of a permanent public collection and served as a touchstone for comparison. The British Museum, for example, was unfavorably compared with the *Muséum* throughout the first half of the nineteenth century. Located at Montagu House, where it had opened to the public in 1759 and remained until 1830, the British Museum was more like the Portland Collection or other large aristocratic collections of the eighteenth century than the more public *Muséum*. The British Museum bird collection was not well kept or well used. It grew at a very slow pace, relying mostly on gifts. Due to lack of space it was unable to accept any major addition, such as the Lever or Bullock collections, and this problem was ostensibly the main reason the government did not purchase those two private museums. Insect pests ravished much of what was owned, and it is little wonder that the Admiralty and other potential sources for acquisitions were reluctant to make contributions, an unfortunate tradition that persisted longer than was actually warranted. This is all not to suggest that the collection was insignificant.[30] Specimens were bought at important auctions (*e.g.*, Bullock's) and the government in 1816 did purchase Montagu's collection, giving the British Museum what was believed to be a more or less complete set of British birds. It is clear from the many citations to British Museum specimens in Latham's writings that the collection was of considerable importance. But, there was no comparison to be made with the French national collection in the first three decades of the nineteenth century. The move in 1830 of part of the bird collection from Montagu House to the nearby new beautiful classical building began a process whereby the British Museum finally emerged as the British national collection, on a par with the French. But that was a number of years away. As late as 1835 the Parliamentary *Report from the Select Committee on the Condition, Management and Affairs of the British Museum* contained much testimony on how the British Museum could become a major national museum, and the *Muséum* was repeatedly cited as an example of what consituted such a museum. Among its more significant aspects the *Report from the Select Committee* noted the importance of having collections maintained by individuals "distinguished in their several branches of science",[31] both for the proper care of them (preservation, labeling, classification), and as an incentive for donors to place their collections in a worthy institution. The *Muséum* figured importantly, as well, in hearings on the reorganization of the Zoology Department into separate divisions; the relationship of a central national museum to smaller collections; the advantage of placing all government-collected specimens in a national collection; and the value of employing "travelling naturalists".[32]

The general tendency in the nineteenth century was for collections to coalesce into a few central public museums, like the *Muséum*. The process, however, was one that was more evident after 1830 than before and also was a tendency partly counteracted by the proliferation of public and private collections.

In Great Britain the establishment of corporate and learned society collections was especially important in the first half of the century for they were the principal loci for the wealth of data that was arriving in ever-increasing amounts. Many of these collections ultimately were incorporated into the British Museum once it became indeed the national museum. For example, one of the most important of collections in Great Britain in the early decades of the nineteenth century belonged to the East India Company, which was practically a government in its own right. It had founded a museum open to the public in 1801 and assembled a major collection of birds which was catalogued by Thomas Horsfield, the keeper of the museum before the company's disbandment and the removal of its collection to the British Museum.[33] The Zoological Society began in the middle 1820's to develop a collection which while it was in existence was also one of the most important ones in Britain. Sir Stanford Raffles (1781–1826) donated his collection to it as did Nicholas Aylward Vigors (1785–1840); John Gould (1804–1881) was curator and taxidermist. Because of its zoological garden, the Zoological Society caught the public's attention and soon became quite fashionable and well supported. Vigors wrote to Charles Lucien Bonaparte in 1828 of the success:

You will be pleased to hear our Zoological Society continues to flourish. Our members advance rapidly, and our attractions in proportion. We are patronized by the most influential part of our community, the fashionable world; – and thus have already become the *rage* of the present day. In this flurry of success we do forget science.[34]

Vigors was overstating the case with respect to the lack of scientific activity. While it was true that the menagerie was what caught the public's attention, it was equally true that this private society became the focus of English zoology and that its museum collection flourished. It, in fact, was the principal recipient of specimens from the government and private expeditions, a situation that was very galling for the staff of the British Museum.[35]

In addition to private societies, universities and local public museums began to play a prominent role among collectors. This was especially the case in Holland and Germany. The Dutch had been avid collectors for a century[36] and their passion was noted by contemporaries. Levaillant in his *Histoire naturelle des oiseaux d'Afrique* wrote that "Holland contains within its small confines perhaps more amateur collectors of all sorts than is to be found in all the rest of Europe together."[37] So it was only to be expected that when a major museum emerged it would be an impressive

one. In 1820 the *Rijksuniversiteit te Leiden* was named the site
for the *Rijksmuseum van Natuurlijke Historie*.[38] The museum was
to be a merging of the university collection, which had recently
been enriched by a sizable collection from Paris in retribution for
the Stadtholder's collection, the government cabinet founded during the brief reign of King Louis Napoleon, and the private collection of Coenraad Jacob Temminck one of the foremost private collectors of the day, and first director of the museum. All totaled,
the museum began its history with almost six thousand mounted
specimens! Since it was the repository for the specimens brought
back by Reinwardt as well as the *Natuurkundige Commissie* in
the Dutch colonies, it quickly grew and by 1835 was in Gould's
opinion the foremost bird collection in Europe.[39]

The Leyden Museum in many ways was not typical of university collections, for although in principle it was an academic as
well as national museum, in practice it was more a separate government institution.[40] In its early years its budget was separate from
the university's and it strongly resisted the suggestion that it
function as an educational establishment. However, even though
the museum did not participate in the teaching functions of the
university, it was one of its outstanding institutes.

From a modern perspective it is easy to be critical of the
"unscientific approach" of its first director, Temminck, who was
typical of the museum directors of his day. He used duplicates
for exchange rather than for establishing series of specimens; he
was not overly interested in geographical distribution; and he was
rather casual in his treatment of type-specimens. However, his
main goal was the enlargement of the collection through government supported *voyageurs-naturalistes,* purchases, and exchanges.
He not only greatly expanded the *Rijksmuseum van Naturlijke
Historie*, but he also, like the French professors of the *Muséum,*
aided other Dutch museums by sending them his duplicates and
by helping them to organize their material.[41] That his goals were

not the same as those of the directors who came after him, for example his successor Herman Schlegel (1804–1884), reflects the time during which Temminck lived, a period that bridged the gulf between the amateur *cabinets d'histoire naturelle* of the eighteenth century and the great scientific institutions of the latter half of the nineteenth century.

It was in Germany that the university museum most fully developed between 1800 and 1830. The outstanding collection, of course, was at the new royally endowed *Universität zu Berlin* which was founded in 1810.[42] Johann Carl Wilhelm Illiger (1775– 1813), a brilliant, but unfortunately short-lived, naturalist was the first curator. The bird collection had at its core the Pallas collection and the cabinet of Count Johann Centurius von Hoffmannsegg, mentioned in the previous chapter as one of a few avid collectors who had the resources to send his own collector into the field. About the time of the founding of the museum Hoffmannsegg had been receiving large shipments from correspondents in South America, and these were supplemented when Sieber returned, shortly after the museum was established, with an immense collection from South America plus specimens from Australia and North America that he had acquired in London by exchange. When Martin Heinrich Karl Lichtenstein (1780–1857) assumed directorship in 1813 the museum had already two thousand specimens representing nine hundred species.[43] Generous financing by the Prussian government allowed Lichtenstein to expand the collection enormously by sending expeditions to various parts of the globe.[44]

The Berlin *Zoologische Museum* served as a model for other new German museums, both university and state museums. Although Prussia and the other German states did not have a colonial empire or extensive maritime resources, many of their museums followed the lead set by Paris, London, and Leyden and assembled large collections of exotics. They were able to do this due to the zeal of the several collectors they sent out, generous donations of private collections, and by exchanges and purchases. Like the Berlin museum these collections were directed by knowledgeable men who appreciated the significance of their material and helped bring Germany to a position of prominence in ornithology. Among the better known were Johann Jakob Kaup (1803–1873) at the grand duke's *Naturalien-Cabinet* in Darmstadt, Philipp Jakob Cretzschmar (1786–1845) at the museum of the *Senckenbergische Naturforschende Gesellschaft* in Frankfurt am Main, Christian Ludwig Nitzsch (1782–1837) at the *Zoologische Museum* at the *Universität Halle*, Heinrich Gottlieb Ludwig Reichenbach (1793–1879) of the *Königlichen Naturalienkabinett* in Dresden, and Johann Wagler (1800–1832) at the *Zoologische Museum* at the new university in Munich,[45] the *Ludwig-Maximilians Universität zu München*.

Great Britain, France, Holland, and Germany had the leading ornithology collections by 1830. They were not, however, a cartel. In Sweden, for example, the *Naturhistoriska Riksmuseum* in Stockholm had a major collection of Swedish fauna assembled by Sven Nilsson plus many exotics. The *Zoologisk Museum* in Copenhagen had grown from a small cabinet to a large collection by absorbing local collections and receiving a wealth of material from travelers like Anders Sparrman (1748–1820). F. A. Bonelli had built a major collection at the *Museo zoologico* in Turino, and the *Kais.-Kön. Hof-Naturalien–Cabinet* in Vienna was collecting on a major scale.[46]

Although the tendency in the nineteenth century was towards the coalescence of collections and the rise of institutional collections, one should not think that private collections were unimportant in the first thirty years of the century. Coenraad Jacob Temminck's collection (before its incorporation into the *Rijksmuseum van Natuurlijke Historie*) was famous throughout Europe.[47] He had had the good fortune of having a father who was not only wealthy and influential (treasurer of the East India Company), but who himself had also collected birds and was on friendly terms with other major collectors in Holland. In Great Britain there were the well-known private collections of William Yarrel (1784–1856) and Lord Stanley (Edward Smith Stanley, 13th Earl of Derby, 1775–1851). Yarrell's collection of British birds was the basis for his *A History of British Birds* (1843), an avifauna that replaced Montagu's as a standard reference. The collection was admired by Bewick when he visited it in 1828, and Bewick's impression survives in a letter he wrote to Dovaston where he stated that: "I never in my life was so gratified at seeing anything, as I was with Yarrell's Museum — he has left nothing untouched that could assist him in probing ornithological knowledge to the bottom."[48] Lord Stanley's collection at Knowsley was even larger and included an aviary, which at his death contained one thousand, two hundred and seventy-two individuals of three hundred and eighteen species (poultry excluded).[49] The mounted collection contained "many of the original examples from the Leverian and Bullock's Museums, described by Drs. Latham, Shaw, and others; the originals of Salt's Abyssinian Collection described by the Earl of Derby", etc.[50] Lord Stanley, who served as president of the Zoological Society, was quite serious about his museum. The unfinished printed catalogue of the Knowsley Collections states that:

These collections have not been formed for the gratification of a whim, for the mere sake of collecting, or for ostentatious display, but have been brought together by his Lordship, at a vast expense, solely for the purposes of recreation and study, — an assertion which may be easily verified by an examination of the under surface of any one or all of the stands upon which the specimens are mounted, where, in his own handwriting will be found noted the sex, when known; the scientific and trivial names of the respective species; a reference to the works of Latham or some other author in which they were described; the colour of the eyes, cere, legs, &c.; the date of purchase; of whom purchased; the date of the death of those which had lived in the Menagerie or Aviaries; and a number evidently intended to refer to a Catalogue, a desideratum which I hope to be enabled to supply.[51]

In France, although the *Muséum* tended to overshadow others, some private collections were notable. Baron Noël-Frédéric-Armand-André de La Fresnaye (1783—1861) in Normandy amased a collection which at his death contained almost nine thousand specimens, over seven hundred of which were new species, many described by him. His collection held many of the specimens brought back by Alcide d'Orbigny (1802—1857) from South America, as well as many specimens purchased from the Maison Verreaux and other dealers.[52] La Fresnaye's collection was known for its exotics as was the museum of Baron Meiffren Langier de Chartrouse,[53] who collaborated with Temminck on the *Nouveau recueil de planches coloriées d'oiseaux* (1820—1839), and the collection of Victor Masséna, Duc de Rivoli, Prince d'Esling (1758—1817).[54] Other French collectors such as Jean-Jules Duchesne de Lamotte (1786—1860) and Jean Crespon (1797—1857) made significant collections of regional species.[55]

In Germany, in addition to the institutional museums there were several important private collections. We have referred already to the splendid museum of Count von Hoffmannsegg and the Naumann collection. There was also that of Prince Maximilian of Wied-Neuwied who had himself collected in North and South America.[56] One should also note a German collection that would have been among the most important of the early part of the century had it survived. The destruction of collections by insect pests was a major problem in the eighteenth century, however it was not the only one; so were invading armies. In the same year as the French were reorganizing their *Jardin du Roi* into the *Muséum National d'Histoire naturelle,* their citizen army burned the castle of Duke Carl II von Pfalz-Zweibrücken (1746—1796) in the Palatinate. Duke Carl II had purchased the major collections of

Bécoeur and Mauduyt de la Varenne and they would have served as the core of a splendid private museum.[57]

One could easily extend this discussion of private and institutional museums by considering the United States, India, South Africa, Australia, etc.[58] Indeed throughout the Western world, and the regions under its control, museums of natural history existed. What is especially striking about the natural history collections as they developed from 1800 to 1830 (especially after 1815), be they national, local, public or private, is that unlike the museums of the previous century, the majority of these were serious working collections, and those museums open to the public, *e.g.*, the *Muséum* in Paris, were a far cry from the Leverian Museum or the cabinets of which Pallas complained. The nineteenth-century collections were more serious for a number of reasons. Partly fashion had run its course, and the political events of the 1790's and early 1800's were not conducive to the amateur collector. A more important factor was the sheer size of the new collections, which completely overshadowed the amateur collector. A gentleman could afford a few dozen birds; a nobleman several hundred. But the scale of collecting undertaken by the French, Prussians, and Dutch relied on a more serious commitment. By twentieth-century standards, of course, these collections appear very backward and amateurish. Birds were mounted and thus exposed to the damaging influences of light, heat, and air. Curators had little or no formal training and appear to us to be more interested in acquiring than studying their specimens. The full significance of type-specimens was not fully comprehended and the result was some rather formidable confusion for later ornithologists. Museum directors like Temminck often used their duplicates for exchange or like Lichtenstein sold them to buy others,[59] rather than attempting to collect series of birds. However, in spite of the deficiencies of curators and the obvious weaknesses of collections when compared to directors and collections later in the century or in our own century, the work done between 1800 and 1830 was a turning point in natural history collecting. The sheer wealth of material acquired was staggering by earlier eighteenth-century standards. Collections of the scale that Buffon and Brisson envisioned as something in the distant future could be found throughout the world. An entire natural history business was flourishing, and collections were growing dramatically each year. The collections were not playthings of the aristocracy or of eccentrics, but serious scientific establishments. The importance of these collections must not be underestimated. They housed type-specimens and the material for scores of monographs. They led to the production of catalogues and raised interesting questions for classification. They were, in a word, the center of ornithological activity and determined the direction that study took.

Given the diversity of collectors and collections it is difficult to perceive a simple cause for the sudden proliferation and development of natural history collections. An enormous potential not available before existed in the navigational and commercial expeditions associated with a new wave of colonialization, and this potential could be exploited due to increased funding which was used in different ways and for different motives. Wealthy individuals like Prince Maximillian or Lord Stanley could acquire hitherto unimaginably rich collections. The French government during the revolution decided not to dismantle the *Jardin du Roi*, as it had done with most of the other scientific institutions of the *ancien régime*, and instead accepted the proposed reorganization to a national museum of natural history; one that would encourage the solution of practical problems; one that would provide lectures to the public; and one that would reflect the glory of France as the leader in the world of ideas. In Berlin the creation of a new university signaled the start of a monumental reform of German universities and coincided with the interests of Hoffmannsegg and the chancellor Wilhelm von Humboldt which resulted in the

Zoologisch Museum. In England the lack of support given to the British Museum led to the establishment of the museum of the Zoological Society, which because of its tie to the practical study of acclimatization of foreign species to England and its extremely popular zoo, permitted it soon to surpass the British Museum, and indeed most other collections in Europe. Although still accessible to a small percentage of the total population, museums by 1830 were a nexus where a number of interests fruitfully crossed. They were supported by those of the evangelical persuasion because of their edifying value, for the study of nature was a path to God, a position eloquently expressed by Paley's *Natural Theology*[60] (1801). They were supported as well by freethinkers, who saw these secular institutions devoted to the study of nature the sign of the triumph of rationality over superstition, and science over theology. They were also supported by a rising middle class for whom cultivation of the sciences was a means of legitimatization. Museums were reflections of empire, of local or civic pride, of cultural respectability, and of academic forwardness. The fortunate conjunction of so many diverse interests prepared the background for the take off of natural history collections after 1830, after which time collections grew even more rapidly and central national collections absorbed many private museums. The growth of collections up to 1830 also determined the ornithology produced up to that date, for just as Brisson's and Buffon's work had been tied to specific collections, as we will see in what follows, the work done between 1780 and 1830 was also similarly closely tied to ornithological collections.

CHAPTER V

ORNITHOLOGICAL PUBLICATIONS: 1780–1800

The accelerating influx of ornithological data after Brisson and Buffon's time was bound to have a profound impact on the nature of ornithology, although the manifestation of this influence was gradual and therefore undramatic. Natural history collections until roughly the turn of the century retained, for the most part their amateur curiosity cabinet status, and one finds in the ornithological literature a reflection of the patterns established by Brisson and Buffon. Brisson, as was described in chapter two, directed his attention to classification based on external morphological features, whereas Buffon attempted a broader natural history of birds. In a sense they epitomize two traditions in zoology; that of the taxonomist and that of the naturalist. This is not to suggest that zoologists necessarily have followed one approach exclusively; they have most often participated in both traditions, in differing degrees, or in others as well. In the ornithology of the late eighteenth century, the direction given by Brisson and Buffon is particularly evident for most of the literature consists of new editions, popularizations, attempts at updating or detailed investigations of a more narrow focus that supplemented Brisson or Buffon's ornithology.

The high literary quality and general appeal of Buffon's *Histoire naturelle des oiseaux* led to its popular success and consequently to the subsequent publication of numerous pirated editions, anthologies, and translations.[1] Buffon himself had published his work in four different formats, of varying price, in order to attract a wider range of prospective purchasers, and publishers were quick to see the potential in his writings. By 1800 Buffon could be read

68

in all major European languages, was excerpted in encyclopedias, appeared in children's versions, and was imitated abroad. Not that Brisson was ignored. Although his six volumes of morphological descriptions, written more in the style of a reference book than a literary one, never attracted a popular audience, they were widely used. Oliver Goldsmith (1728–1774), for example, relied on Brisson for the ornithological section of the first edition of his *Animated Nature*[2] — one of the most popular books of the century — and synopses of Brisson's method could be found in most popular ornithology works or in encyclopedias.

The proliferation and popularization of Brisson and Buffon's ornithology reflect the growing audience for natural history in the late eighteenth century, a trend that can also be detected in the increased amount of natural history to be found in the periodical literature, in the number of scientific dictionaries and encyclopedias published, and in the emphasis on nature in the arts. We can gauge Brisson and Buffon's importance for this audience by looking at their status in the major popular scientific reference works of the period. The Paris publisher Charles-Joseph Panckoucke (1736–1798) began printing in 1782 a famous encyclopedia entitled *Encyclopédie méthodique* which was an expanded and updated version of D'Alembert and Diderot's popular work. The *Encyclopédie méthodique* became quite celebrated, mostly because of the high caliber of its contributors which included Daubenton, Vicq d'Azyr, and Lamarck. The ornithological section was done by Pierre-Jean-Etienne Mauduyt de la Varenne (1730–1792), who owned one of the largest bird collections in Paris. Both Brisson and Buffon had used his collection for their ornithologies, and Mauduyt, in turn, used their ornithologies for his encyclopedia. Basically he relied on Brisson's method, which he praised as useful and easy, but which he freely modified where he felt Brisson was incorrect or uninformed. As for Buffon, he wrote: "Monsieur Buffon is, then, the first to have given us, actually, a general history of birds."[3] Mauduyt extensively utilized and praised Buffon's synonyms, *Planches enluminées*, and natural histories. Mauduyt began publishing his *Ornithologie* before Buffon had completed his *Histoire naturelle des oiseaux* and in most ways was inferior, being basically a compilation, mostly taken from Brisson and Buffon.

This is not to suggest that all late eighteenth-century ornithology was derivative. The work of John Latham, for instance, seriously attempted to update and to incorporate the material newly available in the 1780's and 90's into ornithology. Latham in the preface to his *A General Synopsis of Birds* (1781) wrote:

The intent of the following sheets is to give, as far as may be, a concise account of all the Birds hitherto known; nothing having been done in this way, as a general work, in the English Language, of late years. In other countries, however, it has been paid more attention to; witness that valuable work of *M. Brisson*, who has brought down his account to the year 1760, when he published his *Ornithology*. That great and able Naturalist *M. de Buffon* is likewise proceeding fast with a grand work on the same subject [seven of the nine volumes were done] which when finished on the extensive plan that he has chalked out for himself, will do him much honour . . . . In this work of *M. de Buffon*, not only everything which has been treated of before is properly noticed, and the many contradictions of various authors reconciled, but many new subjects have been added, rendering it a valuable work.[4]

Latham paid his respects to Brisson and Buffon, and fully utilized their work, however, he looked elsewhere for his method of organization and relied as much as possible on personal observation in the collections available to him. He wrote that he would be able to expand the empirical base of ornithology,

. . . from the numerous collections in Natural History, which have been formed of late years in England, and in which, in course, a multitude of new subjects have been introduced from various parts of the world; but more especially within these few years, from the indefatigable researches of those who have made so great discoveries in the Southern Ocean.[5]

In this latter remark he was alluding primarily to the specimens and drawings brought back on Captain James Cook's (1728–1779) voyages (1768–1771, 1772–1775, 1776–1780), which were dispersed among a number of collections.[6] Since Latham informed his reader of the location of specimens used for his descriptions, it is clear that he had access to the Cook material and other new material in the major collections of his time: Banks, Tunstall, Portland, the Royal Society, British Museum, Lever, etc.

Latham continued to revise, supplement, and expand his ornithology so that it included everything known, utilizing collections and other published works. His zeal led to his reputation as the most "renowned of the British ornithologists of the eighteenth century"[7] and the grand old man of British ornithology until 1837! It is not surprising, then, that he often had first access to new material, such as the notebooks of birds of New South Wales that the collector Aylmer Bourke Lambert (1761–1842) had obtained from John White's (1756–1832) voyage, much of the Cook data, and General Hardwicke's Indian bird material.[8]

Latham's ornithology was closer to Brisson's than Buffon's, in that he was primarily interested in recording and identifying the known birds of his time. He worked from collections and publications and, as might be expected, his emphasis was on external physical characteristics, not general natural history. The collections he utilized contained little information about the habits or habitats of the birds and, as in Brisson's writings, one often finds the phrase "native place unknown". The basic organization of Latham's work, however, did not stem from Brisson. He appreciated Brisson's careful descriptions, and in the preface to the first edition of *A General Synopsis of Birds* he said of Brisson's ornithology:

Whoever has perused this work, will be fully convinced of the accuracy and precision with which this gentleman has treated the subject throughout; and it is but justice here to acknowledge the liberty we have taken with these descriptions, in respect to such Birds as have not fallen under our inspection.[9]

Nonetheless, when it came to classification Latham relied on Ray for his basic divisions and on Linnaeus for lower divisions. As was mentioned in a previous chapter, Ray's was the best ornithology available in the middle of the eighteenth century, and it had a deep influence on English natural history. Linnaeus's influence is more problematic to understand, for his classification of birds was not distinguished. His interest in birds was secondary. He did not have access to a major bird collection, nor does it appear that he was as conversant with the published literature as was Brisson or Buffon. In successive editions of the *Systema Naturae*, the only general treatment of birds Linnaeus published, he included numerous newly reported birds, but the basic organization into six orders remained the same, although he moved some genera from one order to another. In his last edition (the twelfth), published in 1766, Linnaeus recognized seventy-eight genera and nine hundred and thirty-one species, a good number having been adopted from Brisson. (The 1758 edition had sixty-three genera and five hundred and forty-five species.)[10] Linnaeus's general popularity in the eighteenth century, everywhere except in France where the strong traditions at the *Jardin du Roi* kept him at bay, has to be understood by reference to his reputation as the Prince of Taxonomists.[11] He was the recognized authority in plant systematics, and his methods and writings covered the entire range of living objects. His reputation outside botany in large part rested on his introduction and consistent use of a binominal nomenclature, which brought some order to the confusion in nomenclature, and on his *Systema Naturae per Regna Tria Naturae*, which attempted to place rationally and clearly characterize all known forms in natural history, and which he frequently revised (twelve editions between 1735–1766). The *Systema Naturae* soon set the standard for classification, for it was convenient, harmonious, and useful; nothing was comparable. In England and many parts of the continent, the tradition established by Linnaeus in systematics persisted, and in a very modified form still does. Latham, who was influential in founding the Linnean Society of London, openly acknowledged his debt to Linnaeus. He wrote in 1821 to James Edward Smith (1759–1828), the founder of the Linnean Society,

When I took up *Linnaeus* I thought I had sufficiently trodden on his Heels, in adding to his Genera (79 in all) – 12 new ones – & in my present Work will be 8 more, making in all but 20 new – in the whole 111 – & this has been in great measure owing to new Subjects unknown to Linnaeus but on going [?] over Mr. *Temminck* I find *201 Genera*, and thou He may be right in many, He is no [sic] so happy in all, besides I perceive that the modern Systematists wish to lose sight of Linnaeus as much as may be, but neither you nor I can allow of this. Alter & improve if you please, but still have your grand Distributor in View." [12]

Not that Latham was a mere epigone of Linnaeus. He used Linnaeus's genera, and in some of his works he imitated Linnaeus's style of short diagnoses, however, he freely disagreed with Linnaeus's arrangement and modified it accordingly. He also, in his *General Synposis*, used English names for birds rather than Latin as required by the Linnean system, and he persisted in all his work to use Ray for his general divisions into land and water birds.

Latham's work is instructive as a reflection of the best ornithological work in the late eighteenth century. Even though he lived to a good old age and was active well into the nineteenth century, he was in outlook and style an ornithologist of an earlier period. Dr. Latham was a practicing medical doctor, who did his research and writing in his free time. Although he was an important member of the scientific community, had access to major English collections, and corresponded with practically all the important figures in the natural history community, when it came to technical ornithology, he was working mostly in isolation. He, like Edwards before him, etched his own copper plates and was severely taxed with all the complications of trying to publish his bird books, which appeared in small editions of a few hundred copies.

The audience Latham wrote for was a general natural history audience, for there was no recognizable ornithological group or set of readers. His goal was to keep up with the expanding knowledge of birds, and in this attempt he was continually frustrated; supplement followed supplement, and he forever was telling his reader that he had more material that he would publish in due course. The job was, of course, impossible for him. In his younger years he could not work on ornithology full time, and after he retired to Romsey in 1796, he lived a distance from the major English collections. In addition, there were the European collections, the contents of which he knew only from printed sources. As he grew older, material was accumulating too fast for him to keep up with it; and consequently he never completed the projected second edition of his *Index Ornithologius.*[13] The job of cataloguing all the birds and bird names belonged to a younger generation, who would have different opportunities than Latham had. Latham's general histories and indices had kept general ornithology up to date. Eventually, however, due to the flood of material in the first three decades of the 1800's, he fell behind.

Latham's general ornithology in the late eighteenth century was the most extensive attempt to update the works of Brisson and Buffon. There were many other valuable contributions of a more limited scope to ornithology during the same period. Most notable were the regional studies, either European or exotic. As was discussed in a previous chapter these studies varied considerably both in quality, motivation and approach. Their overall contribution is not, however, to be minimized. Gilbert White and Johann Matthaeus Bechstein added from personal observation to the knowledge of their avifauna, as did Molina, Sonnini, and Levaillant on exotics.

The last two decades of the eighteenth century were a vitally important and dramatic time for Western political, economic, and social history. They were equally an exciting period in the physical sciences, arts, and literature. They were, however, rather quiet ornithologically. Brisson in the 60's and Buffon, primarily in the 70's, had succeeded so well in defining and demonstrating ornithology as a branch of natural history worthy of separate treatment that for a period they more than satisfied the need for a new general ornithology. The two of them, of course, were responding to the opportunity afforded them by the existence of some hitherto unimaginably large bird collections and a general taste for natural history in their society. Not many were as well placed as those two, and the only serious general ornithologist until after 1800 was Latham, who drew heavily on his predecessors. The volume of ornithological literature, nevertheless, did increase. Considerable space was devoted to birds in the scientific dictionaries and encyclopedias of the time, and numerous popularizations and translations were published. Regional studies began to appear in numbers and of a quality not found before 1760, and all the major regions of the globe had its documenters. Local avifaunas contained information on new species and new observations on known species. The natural history of some species in addition to their external morphology was occasionally described.

It should be noted that in some ways the printed record is not a complete guide to the study of birds at the end of the eighteenth century because many observations did not get into print immediately. Access to publication was limited. Although popularizations did well, they were hardly the place for recording original work. Books were costly to print, and scientific periodical literature was still in its infancy. There were occasional ornithological articles in the *Transactions of the Linnean Society of London,* Rozier's *Observations sur la physique, sur l'histoire naturelle et sur les arts,* or *Der Naturforscher,* etc., but they were few and far between, and tended to focus on problems of general interest like bird song, the cuckoo's unusual nesting habits, bird lists, and migration. The recording of new species generally appeared in regional studies, in systematic works like those of Latham, or in general classifications like those of Johann Friedrich Gmelin (1748–1804).

Perhaps the most significant aspect of late eighteenth-century ornithology is that the expectations for ornithology were higher than they had been before. Brisson and Buffon in their introductory remarks revealed a painful awareness of the awesome magnitude of their study and the inadequacy of their information. Although excited by the hundreds of new exotic forms they did have, they guessed that there were many more they did not know and were quite frank in stating that what they projected was merely a first small step. Buffon stated in his *Plan de l'ouvrage* which began the first volume of the *Histoire naturelle des oiseaux:*

We are not attempting to give here a history of birds which will be as complete or as detailed as that which was given for the quadrupeds. The work on the quadrupeds, however long and difficult to bring to completion, was not impossible because the number of quadrupeds is barely two hundred . . . . This work [the birds] is the fruit of close to twenty years of study and research. And although during this time we have not neglected anything that could be of instruction to us concerning birds, or could help in procuring all rare species, to the point that we have succeeded in rendering this part of the Royal collection more numerous and more complete than any other collection of the same sort in Europe, nonetheless we must admit that we still lack a rather great number.[14]

Compounding the problem was Buffon's recognition of the sexual, seasonal, life stage, and geographical variation that birds display, plus the seemingly overwhelming practical difficulties of making observations on the life history of beings who fly, migrate, and (according to Buffon) hybridize profusely. Buffon, of course, was using as a measure a complete natural history of birds. To men of more modest ambition – White, Bechstein, Sonnini – the problems in ornithology, although difficult, were finite. Certainly Europe, one of the areas of the globe with the least number of different species, did not seem so overwhelming, especially given the start made by the volumes of Brisson and Buffon with their plates and synonyms. George Montagu barely a decade after the turn of the century felt that British ornithology was about complete. Although that actually was not the case,[15] it illustrates the point that ornithologists did not feel overwhelmed, and that they had greater expectations of an end in sight. We will see that the deluge of data in the 1820's did not really dampen that view, although the quantity of information was staggering.

The popular success of Buffon's ornithology and the subsequent successes of books like Bewick's *British Birds*, Bechstein's *Gemeinützige Naturgeschichte Deutschlands*, and Levaillant's *Histoire naturelle des oiseaux d'Afrique* demonstrated that an audience for ornithological works existed. It was hardly an ideal audience. Latham's serious surveys were limited to small editions and financially almost ruined him. A work devoted exclusively to birds had to appeal to a general reader. There was no recognizable discipline of ornithology in which a recognized group of experts worked on an agreed upon set of problems using accepted methods and united by some general common goals. In this disciplinary sense, there were no ornithologists. Those who were so-called wrote on birds, but did not do so for other ornithologists. The distinction is not between an amateur and professional, but rather between a specialist and non-specialist. The "ornithologists" before 1800 were naturalists who wrote about birds, but who often also did work in other departments of natural history or science. An ornithology book, of the highest level, was intended for those with a serious interest in natural history, or more commonly, a general reader with a general interest in natural history. Even the most professional institution, the *Muséum*, although it had a chair for mammals and birds, did not have a bird specialist. The few exceptions, the men who found themselves devoted singly to the study of birds, were, like Latham, rather isolated. We should not make too much of this. The amount of material in collections, the quantity and quality of publications, and the number of people involved in the study of birds were greater than had ever been seen. But it was merely the beginning. Although ornithology had not changed significantly from the patterns set by Brisson and Buffon, it had been popularized and extended. Its empirical base had grown, and an audience of some sort existed. The next generation had something on which to build, and build they did.

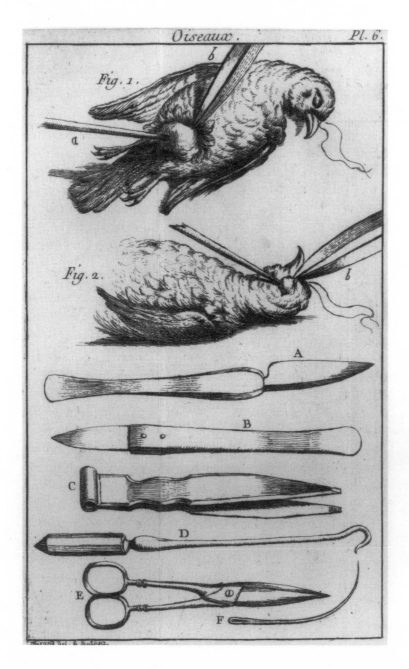

Illustration 4. Taxidermic techniques and instruments proved critical in the development of natural history collections. This plate—illustrating how to stuff a birdskin and place an artificial eye in it, along with some basic instruments—is from one of the first manuals on preparing natural history specimens, Étienne François Turgot, *Mémoire Instructif sur la manière de rassembler, de préparer, de conserver, et d'envoyer les diverses curiosités d'Histoire Naturelle*, Paris, Bruyset, 1758, Plate 6.

Common Osprey. Fish Hawk.

Drawn from Nature by J.J.Audubon F.R.S.F.L.S.   Lith⁴ Printed & Col⁴ by J.T.Bowen. Philad⁴

Illustration 5. John James Audubon is one of the most famous illustrators of natural history subjects. This dramatic plate of the common osprey appeared in his *The Birds of America*, New York and Philadelphia, Audubon, 1840, Vol. 1, plate 15.

Illustration 6. In the nineteenth century, William Yarrell had one of the best-known ornithological collections in England. It was the basis of his multi-volumed *A History of British Birds*, London, Van Voorst, 1843, which was a standard reference for many years. This illustration of a Dunlin chick (Vol. 3, p. 19) reflected the attention given by naturalists to different life stages of birds.

Illustration 7. British naturalists used their hunting skills to collect birds. Often, however, their enthusiasm meant that many specimens were destroyed. This illustration from William Yarrell, *A History of British Birds*, London, Van Voorst, 1843, Vol. 3, p. 528, depicts an Englishman collecting birds in Iceland. Yarrell himself gave up shooting once he became interested in natural history.

Illustration 8. Since some birds are nocturnal, naturalists devised methods to snare them. This picture from William Yarrell, *A History of British Birds*, London, Van Voorst, Vol. 1, 1843, p. 478, shows a method of catching birds at night.

INSESSORES.
CONIROSTRES.

FRINGILLIDÆ.

THE GREENFINCH,

OR GREEN GROSBEAK.

Illustration 9. This greenfinch typified the illustrations that made Yarrell's work popular (William Yarrell, *A History of British Birds*, London, Van Voorst, 1843, Vol. 1, p. 479.)

CHAPTER VI

FOCUS ON CLASSIFICATION: ORNITHOLOGY
1800–1820

Brisson and Buffon had helped to bring ornithology out of a general cultural context, which included everything from cookbooks to encyclopedias, and implicitly defined it as the scientific study of birds. In so doing they emphasized external morphology, natural history, iconography, nomenclature and classification. Brisson's ornithology, conceived from the vantage point of a collection catalogue, focused primarily on external morphology and classification, whereas Buffon's nine volumes were part of a general natural history, which originally was intended to cover all of its branches and consequently attempted a much broader sweep. Brisson and Buffon did not establish a scientific discipline, but they did demonstrate an approach to birds that was popularized, copied, and developed. In the two decades immediately following the appearance of Buffon's *Histoire naturelle des oiseaux* the quantity and quality of ornithological publications increased. In part this development had been stimulated by the work that had just been done; in part it was a response to the arrival of exotic material and the potential of regional faunas which made the study exciting and fruitful. By the turn of the century an important set of individuals were developing further the new tradition. These men worked in different countries, in different styles, and in different settings, however they shared a common rigor in method and common interest in a natural classification of birds that justifies placing them together.

Historians of biology often characterize the seventeenth and eighteenth centuries as a time of conflicting positions on the philosophical foundations of classification.[1] At issue was an

79

artificial versus a natural system, with Ray and Linnaeus allegedly supporting the former and Tournefort and Buffon the latter. As, of course, with most dichotomous divisions, close analysis blurs the distinctions. Nonetheless, the general view that a tension existed between proponents of artificial and proponents of natural systems, is useful and not too much of a distortion. In the eighteenth century Linnaeus's artificial classification system, due to its author's unflagging zeal in keeping it as complete as possible, its general harmony and simplicity, and its use of a binomial nomenclature that brought some order to an otherwise chaotic situation, was widely held. Moreover there were several important naturalists who held, on philosophical grounds, that a natural system could not be achieved. Buffon's collaborator on the quadrupeds, Louis-Jean-Marie Daubenton (1716–1800), for example, in the introduction to the natural history portion of the *Encyclopédie méthodique* stated that "All these systematic divisions into orders, classes, and genera depend on the wishes of the naturalist who imagines them. They are not indicated by the nature of things. The same objects are differently classified by different authors, and sometimes by the same author."[2] This nominalist position made sense in an age when collecting was an entertainment and when classifications were necessary as useful keys for organization. The lack of any agreed upon philosophical foundation for a biological classification furthered this trend. Yet by the end of the eighteenth century the search for a natural system was clearly a dominant goal. One can refer to a number of factors responsible for the shift from constructing artificial systems to the search for the natural one. The ascent of the *Muséum*, which had a tradition of natural classification going back to Tournefort and Bernard de Jussieu, and actively kept alive by Buffon, is important to note. Buffon's enormous influence, not only in France but also in Germany cannot be overlooked. Linnaeus's own stress on the possibility of a natural system found roots in Germany, England,

and Scandinavia. The evangelical movement's enthusiasm for natural history as an edifying activity revealing God's plan and hence his omnipotence, foresight, intelligence, and beneficence suggested a search for the order in creation. The dramatic fulfillment of the Newtonian program which synthesized the rationalist and empiricist traditions into a mathematical physics and astronomy, must have been pregnant with the example for natural history that just because ultimate knowledge may not be attainable one need not settle for arbitrary hypothesis or artificial system. One could probably go on listing reasons, all of them valid to one degree or another, and certainly significant in one case or another. But a rather simple and straightforward explanation should not be overlooked: the quantity of material and the quality of examination suggested, empirically, an order. Some naturalists, like Buffon, began their careers with strong nominalist leanings, but after years of study came to believe in a natural order knowable to man. There were others, like the *Naturphilosophen* who began with an acceptance of a knowable order and found their beliefs confirmed by the empirical data. Most naturalists, however, were not very philosophical, and the impressive elucidation of what appeared to be an order in nature gave them the confidence that a natural system would be forthcoming.

Perhaps more than anyone the person who did the most to reinforce this view was Georges Cuvier (1769–1832) whose comparative anatomy was widely perceived to be the tool that would make the study of animals into a rigorous science.[3] Cuvier had come to the *Muséum* in 1795, shortly after its reorganization, to assist Etienne Geoffroy Saint-Hilaire. In two decades his stellar career brought him to a position of social, political, and intellectual preeminence. Cuvier never published a major treatise specifically on birds, as he had done on fossil quadrupeds, on mammals, and on fish, yet his influence was profound. Like Linnaeus, who also never published separately in ornithology, Cuvier constructed a general classification system that had an almost universal impact.[4] The force of the system resulted from its foundation in the science of comparative anatomy, which Cuvier had helped bring to a high level of sophistication, and through which he attempted to uncover the basic laws of biology. Cuvier had formulated anatomical laws that established a science of animal forms and which he believed held the key to a natural classification. With this insight, Cuvier examined the extensive holdings of the *Muséum's* bird collection and constructed an important system. Although his comparative anatomy helped to transform systematics, ironically, Cuvier's own interest in the subject was secondary. Of bird classification, he had written in 1790 to a correspondent:

Since I have seen Brisson's ornithology, I can no longer concern myself with the classification of birds. The work, in my opinion, provides all that one can expect on the subject: 1500 species or varieties are described with the greatest care and more than 300 are engraved on copper. It is true that often simple varieties are raised to the rank of species, but it is the business of naturalists to correct these errors, and Buffon, fortunately, has done it in the majority of cases.[5]

When Cuvier, two decades later, began a reorganization of the bird collection (1811), the *Muséum* possessed four thousand birds[6] and Brisson's *Ornithologie* was considerably out of date. Cuvier's arrangement, which is found in his famous *Le Règne animal* (1817) reflected his approach to classification. His main interest was in understanding the form and function of animal parts, and consequently his systematic works were encyclopedias of all animal forms, rather than keys for identification or literary descriptions. The higher taxonomic levels were more interestingly discussed by Cuvier, for in their characterization he rationalized the groupings based on "conditions of existence", the same fundamental concept that he used for his basic anatomical laws.[7] On lower levels, he was satisfied to utilize arbitrary elements. Cuvier basically stuck to Linnaeus's six orders, with some modifications, although he claimed that his general divisions are indicated by the life history of their members. The significance of Cuvier's ornithology is not so much in the specific ordering of birds, but rather in his elevation of systematics to a more scientific base. By 1800 the professors of the *Muséum* were agreed that comparative anatomy was the key to classification. In their attitude, they were reflecting the maturation of comparative anatomy as a science, and it was Cuvier who best exemplified the power of this new study to uncover the structural relations among animals. With this new tool came the key for a natural system of classification based on empirically discovered relations. The problem for the next several decades was seen by many to be the attempt to work out in finer detail the classification system coming from comparative anatomy.

Cuvier's colleagues at the *Muséum*, for example, although they held the same commitment to comparative anatomy, were not satisfied with his avian system. In particular, his use of characteristics from different organs to define subgroups of orders lacked balance, and his adherence to only six orders placed together some very diverse forms. Henri-Marie Ducrotay de Blainville (1777–1850), who later succeeded his mentor Cuvier in the chair of comparative anatomy at the *Muséum*, thought that by studying the sternum one could construct a more uniform system.[8] A few years later, Etienne Geoffroy Saint-Hilaire proposed the basis of another more uniform classification in his *Système dentaire des mammifères et des oiseaux* (1824).

Although the use of comparative anatomy was suggestive and in its more general conclusions in classification widely accepted, it did not completely dominate avian systematics until after mid-century. What Parisian professors did establish by their work in comparative anatomy, however, was the serious expectation that a natural system could be attained. Not that it was necessarily close to attainment. Louis-Pierre Vieillot (1748–1831), whose long career brought him into contact with Buffon as well as Cuvier, in his article "Ornithologie" published in the *Nouveau dictionnaire d'histoire naturelle* (1818) wrote:

There is no, nor could there be, a perfect one [system of classification] as long as one does not know all the species of birds which are distributed over the globe. Only then would one have one which left nothing to be desired, and this will be the natural system .... However, let us not cease to occupy ourselves with them and let us consider all of our systems, more or less artificial, more or less approaching nature, only as an accumulation of material, the choice of which will be of some utility for the future structure of the science of systems.[9]

Approaches varied. Vieillot used external morphological features. In part, that reflects his interest in exotic ornithology where the material available was suited to that approach. It was Vieillot who described many of the species brought back to the *Muséum* by its *voyageurs-naturalistes* and by government expeditions. He developed his own system which he described in his *Analyse d'une nouvelle ornithologie*, and which he used in the *Nouveau dictionnaire*.[10] Johann Carl Wilhelm Illiger, at the Berlin Zoological Museum, also studied exotic materials and used morphological features for classification. But more in keeping with the tradition of comparative anatomy as established by Johann Friedrich Blumenbach (1752–1840), the "German Cuvier", Illiger tried to compare the totality of characteristics, the total *habitus* as the German tradition named it, for establishing groups. This entailed a very careful comparison of internal and external features and a weighing of more important in favor of less important characters. The result was a system that claimed to be built inductively and hence more naturally than those based on variations of a single structure (*e.g.,* Blainville's system). In practice, Illiger relied heavily on the same standard external morphological characters used by most.[11] Illiger's feel for what we now recognize as natural groups has awakened a modern appreciation of his work, more so than in his own time. However, his painstaking research was not ignored, and his revisions, especially among genera was widely accepted. Illiger was also concerned with standardizing terminology, a concern which infuenced many German naturalists and foreshadowed the move towards standardization a few decades later.[12]

The famous collector, and later director of the Leyden museum, Coenraad Jacob Temminck, was one of the systematists who utilized Illiger's work, although he by no means followed it. Temminck also tried to employ the bird's natural history to establish divisions.[13] In criticism of Vieillot's use of solely external characters he wrote: "The knowledge of habits and of anatomy are two sister sciences, inseparable mates of a good systematic classification."[14]

Temminck, who was the most famous ornithologist of the early nineteenth century, had one of the greatest collections of exotic birds and often traveled to the other major European collections.[15] In his writings, we find evidence not only of one of a number of approaches to systematics, but another feature of this group of naturalists who dominated ornithology in the first decades of the nineteenth century: an increased specialization and rigor. Although Temminck came from a milieu of amateur collectors, he attempted in his published works to raise the level of discourse in ornithology. He dryly noted in the introduction to the second edition of his *Manuel d'Ornithologie; ou Tableau systématique des oiseaux qui se trouvent en Europe* that the quest for the unusual was not necessarily the most profitable path to pursue:

One goes to search in the regions of the torrid zone or to the ice of the poles for material to add to the numerous species already known, by means of which one augments the nomenclature catalogues without any goal useful to science. They are sterile acquisitions, that amateur curiosity collectors esteem, but which will be, for a long time, foreign to the domain of science.[16]

Far from being dazzled, then, by his "ooh aah" collection, Temminck, like Cuvier and Illiger, displayed a desire to study birds in a serious scientific manner. Writing in the very midst of the mania for exotic materials, Temminck did the first important European avifauna. He stressed the need, moreover, for studying many specimens of the same species to better understand variation, distribution, etc., rather than just acquiring more species.

Temminck's increased scientific approach is reflected in his *Histoire naturelle générale des pigeons et des Gallinacés*, a three-volume work on a single group of birds, in which he carefully classified and described all the known species of pigeons and gallinaceous birds. This work of Temminck's was one of the earliest ornithological monographs, a genre that was to become one of the distinguishing features of nineteenth-century ornithology and an index to increased specialization. Alfred Newton, in the introduction to his *Dictionary of Ornithology*, wrote:

On reviewing the progress of ornithology since the end of the last century, the first thing that will strike us is the fact that general works, though still undertaken, have become proportionally fewer, and such as exist are apt to consist of mere explanations of systematic methods that had already been more or less fully propounded, while special works, whether relating to the ornithic portion of the Fauna of any particular country, or limited to certain groups of Birds — works to which of late years the name of "Monograph" has become wholly restricted — have become far more numerous.[17]

Monographs on birds first appeared at the very end of the eighteenth and beginning of the nineteenth centuries. Their origin was closely linked to artistic and iconographic publications. J. B. Audebert (1759–1800) and Vieillot's celebrated *Histoire naturelle et générale des Colibris, oiseaux-mouches, jacamars et promerops* is typical of these works which began a period of fashionable luxury bird books, occasioned by improved techniques of printed color-engravings.[18] Audebert was a pioneer in applying the new techniques in zoology, and his work, carried on by Vieillot after

Audebert's death, was a *tour-de-force* of what has become known as the "golden age of ornithological iconography in France".[19] Birds were popular subjects for illustration because of their brilliant plummage, which the artists attempted to capture by printing with gold and oil colors. There was no attempt in the *Oiseaux dorés* as these volumes by Audebert and Vieillot were called, to provide a scientific classification of exotic species. The preface stated quite frankly that: "Since our sole goal is to make known the Hummingbirds by illustrations done by a new and more exact technique than ever used to date, our drawings should do nothing to change nomenclature. We retain the French names given by Buffon and the Latin ones of Linnaeus."[20] Audebert also collaborated with Levaillant in printing (with Temminck's patronage) the *Histoire naturelle des oiseaux d'Afrique* (1796–1802), a more scientific work with splendid plates. After Audebert's death, Levaillant, like Vieillot, continued to publish luxurious ornithology books. He worked with the well-known Jacques Barraband (1767–1809) and with other artists to produce some of the finest works in the genre. The most beautiful, and according to the art historian Ronsil "perhaps even the most beautiful of the ornithological works of the beginning of the nineteenth century",[21] was the *Histoire Naturelle des Perroquets* (1801– 1805).

The splendid French ornithology books of the early nineteenth century were important for iconography and for popularizing the study of birds. A few dealt with a particular group or a particular region and can be considered protomonographs.[22] The closest to a monograph is the *Histoire naturelle des Tangaras, des Manakins et des Todiers* (1805) by Anselme-Gaëtan Desmarest (1784–1838), which in addition to the seventy-two lovely color plates engraved from the drawings of the popular artist and former student of Barraband, Pauline de Courcelles (1781–1851), attempted to bring order to this group and contains a sophisticated discussion on the principles of nomenclature.

Temminck began to publish his monograph with the collaboration of Mlle Pauline de Courcelles, by then Mme Knip, in 1809.
The work, as most in its day, was published in installments. In
1811 after the ninth livraison Mme Knip changed the title page
to make herself appear the main author and made textual deletions
wholly unacceptable to Temminck. The collaboration broke down
amid mutual recriminations, and Temminck went ahead and
published his *Histoire naturelle des pigeons et des Gallinacés*
(1813–1815), a work much less luxurious, but an extensive one
that deserves the description he gave to it, a monograph.[23]

Before 1830 the number of monographs produced were few,
however, their appearance at all suggests the extent to which
collections had grown and the extent to which some naturalists
had specialized. The favorable public reception of specialized bird
books was also important in demonstrating the existence of a
potential, albeit elusive, audience for such works.

Classification, in the forms of encyclopedias, avifaunas, iconography, monographs, and general systematic works, dominated
ornithology in the first two decades of the nineteenth century.
What of the natural history of birds as exemplified and urged by
Buffon? The observation and description of the life history and
behavior of a bird species had become a minor tradition confined
to sections of local faunas. Even in these natural history was
generally subservient to classification, and they tended more and
more as the century progressed to become strictly bird lists.
Popular bird books, which catered to a more general public, often
focused attention on the life history of birds, but these accounts
were not original and added little to the knowledge of natural
history. After mid-century, the volume of this minor tradition
expanded greatly due to the establishment of field clubs and
local natural history societies, but even then it continued to be
secondary to what was perceived to be the central and defining
task of ornithology: classification.

Ornithology focused on classification and the significance of this preoccupation should not be minimized. Taxonomy has often been disparaged; it is often considered to be merely a primitive stage in the growth of science. Lord Kelvin, for example, in his patronizing comments on non-mathematical science stated that:

... when you can measure what you are speaking about, and express it in numbers, you know something about it; but when you cannot measure it, when you cannot express it in numbers, your knowledge is of a meager and unsatisfactory kind: it may be the beginning of knowledge, but you have scarcely, in your thoughts, advanced to the state of *science*.[24]

This attitude completely fails to appreciate the importance of nineteenth-century biology as well as hopelessly confuses the goals and methods of different disciplines. The significance of the focus on classification in ornithology is two-fold. By the end of the eighteenth century and beginning of the nineteenth century the goal of classification was the construction of a natural system. This goal reflected an attempt on the part of naturalists to uncover an existing order in nature; or to put it in terms which even Kelvin would have understood, to discover the laws of nature. Classification, then, was not solely the pragmatic ordering of data for convenience, but could claim to be a serious scientific endeavor. The development of classification also promoted an intellectual specialization which led to the emergence of ornithology, as well as other zoological disciplines. The increased empirical data, and the increased seriousness with which classification was pursued, made such specialization inevitable. Classification, then, in addition to raising the scientific importance of the study of birds by focusing on the search for a natural system, hence an order in nature, defined the emergence of ornithology for the simple reason that it divided nature into large groups, and one of the largest, most interesting, and most natural is the birds.

Ornithologists, moreover, were self-conscious of the scientific rigor of classification which set it appart from less "serious" endeavors. Temminck, for example, one of the key figures in the transition of the study of birds from a sub-branch of natural history to a scientific discipline, perhaps misguidedly and perhaps influenced by his experience with Mme Knip, held the fashionable bird art books of the period in contempt. He wrote that:

It is no longer the love of the science which guides the writer. It is rather a sordid interest that governs his pen, which is sold to a book dealer or draughtsman, who, destitute of the slightest knowledge in natural history and without any guidance, haphazardly makes some drawings, often from worthless remains [bird skins], the degree of deterioration of which he is ignorant, and orders the text from the lowest bidder, thereby producing a work which disgraces both science and the author.[25]

Although the increased rigor of the study of birds and the specialization that demarcated ornithology as a separate department of natural history was evident in the first two decades of the nineteenth century, it would be inaccurate to speak of the discipline of ornithology coming into being by then. For a discipline to be present there must be a sizable number of specialists, in contact with one another, and with common programs and goals. The few individuals between 1800–1820 who focused extensively on birds were widely scattered. They did correspond but their correspondence was irregular and was not exclusively devoted to ornithology, for with the exception of Temminck and Vieillot, most "ornithologists" also ranged over a wide area of natural history. For example, Illiger spent as much time studying entomology as ornithology. The paucity of monographs or local faunas that were strictly avifaunas, is in many ways a good index of limited disciplinary activity for their production requires a single-minded effort which was rare at the time. Funding was equally scarce. Illiger had the good fortune to have a wealthy and knowledgeable patron, the Count Hoffmannsegg; Temminck had independent means; and Levaillant profited from fashion and his lively writing style. Vieillot, however, ruined himself financially. Generally it was a fragile existence. Those few who were able to support themselves by their work rarely were able to do so by studying birds exclusively. The naturalists who did significant work on birds constituted a very small and mixed lot, with diverse backgrounds, and in different settings. However, if they do not show enough family resemblance to be labeled a new generation or a new discipline, they do reflect the increased interest in natural history that made possible an audience for serious writing, the supplanting of the curiosity collector by a more scientific collector, and the generally more rigorous goals of natural history. In the subsequent two decades, the tendencies noticeable between 1800 and 1820 more fully demonstrated their potential and one can legitimately speak of the discipline of ornithology.

CHAPTER VII

THE EMERGENCE OF A DISCIPLINE:
ORNITHOLOGY 1820–1850

European society was fundamentally altered by the combined impact of social and economic revolutions in the late eighteenth century. By 1830 the effect of the ramifications of these two movements was evident. E. J. Hobsbawm wrote that:

Whichever aspect of social life we survey, 1830 marks a turning-point in it . . . in the history of human migrations, social and geographical, in that of the arts and or ideology, it appears with equal prominence.[1]

He could have added to his list the world of science which was transformed along with all the other institutions of nineteenth-century Europe. Documenting the influence of social and economic change on the history of science, however, is a difficult task. In the case of the emergence of ornithology there are some obvious manifestations of this interaction. One has already been mentioned in chapter three: the effect of the enormous colonial expansion that was linked to commercial expansion. National interests were clearly served by government-sponsored scientific expeditions between 1815–1830. After then the "heroic age" was over and maritime activity settled into more mundane activity.[2] John Dunmore, in his excellent study of the French exploration of the Pacific, comments that after Dumont d'Urville's first voyage (1829):

Exploration for its own sake, for science, for the advancement of geographical knowledge and of navigation, played a continually decreasing role. Scientific explorations — indeed, expeditions of any type — became less frequent .... The rise of the propertied middle class, which had already made its power felt under Charles X, was completed under Louis-Philippe.

Commercial activity in the Pacific after 1826 reflected the energy and determination of the mercantile class.[3]

Not that the number of bird specimens coming back to Europe decreased. Far from it; the number of specimens increased exponentially! Government expeditions were not launched to search for new continents but numerous ships were sent to trade and to chart coasts. The British ships *Beagle* (1832–1836), *Erebus* and *Terror* (1839–1843), and the French *la Favorite* (1830–1832), *la Vénus* (1838–1839), and *l'Astrolabe* and *la Zélée* (1837–1840) brought back major collections of great scientific importance. *Voyageurs-naturalistes*, captalizing on established settlements and trade routes, penetrated into lesser known parts, and enough collections and museums existed in Europe to support major commercial collectors like the Maison Verreaux. Pierre-Jules Verreaux (1807–1873) and his brothers Jean-Baptiste-Edouard Verreaux (1810–1868) and Joseph-Alexis Verreaux († 1868) worked extensively in the Cape Colony where they obtained enough material by 1834 to re-establish and supply the family natural history business and soon make it the foremost commercial natural history establishment in the world.[4] Jules also collected with government support for five years in Australia and New Zealand, returning with nearly twelve thousand natural history specimens. One might think that a commercial establishment like the Maison Verreaux would be solely interested in new species and "showy" specimens. The Verreaux brothers, however, were naturalists themselves (like Leadbeater in London) and knew what was scientifically interesting. Their experience in the field also made them sensitive to the value of field observations, and Jules Verreaux was in the minority tradition that stressed the importance of natural history for ornithology. In a notebook of his, after recording some interesting nesting habits, he commented:

I believe that very unfortunately we do not have in all Europe a single scientist who is in this position [of being interested in writing a monograph on the group using his information]; generally scientists know nothing touching on the habits [of birds] which, I cannot tire of repeating, are for me the base of any lasting and stable structure one would build in science. I hope very much, if my voyage ends as I desire, to be able to spread this love, which has enchanted me all my life and to show in an incontestable manner the necessity of training our young scientists to follow such sorts of studies so that in a few years we will see this science receive all the merit that it should occupy in the civilized world, as in Europe we have the pretension of so thinking of ourselves. My only fear, however, resides in the great difficulties that I think I will encounter among our scientists, especially those at the top of this science.[5]

The success of the Maison Verreaux and Leadbeater & Son, which catered to a clientele of serious naturalists, in addition to reflecting the increased volume of specimens available in the nineteenth century, also reflected the general growth and diversification of interest in natural history. The *cabinet d'histoire naturelle* as a fashionable attraction was a part of aristocratic culture. It continued to exist throughout the nineteenth century but its importance was eclipsed by large, serious, working collections, increasingly public, and by a number of local museums that developed toward the middle of the century. Although the large scale proliferation of provincial societies and museums belonged to the second half of the century, by mid-century a noticeable number existed. In Great Britain, for example, provincial societies established museums and popularized the study of birds.[6] The Rev. A. Hume, in his book, *The Learned Societies and Printing Clubs of the United Kingdom* (1847), noted that "one of the first objects in the smallest provincial town where such a society can be organized, is to procure a Museum".[7] These museums usually attempted to stress local species and often employed a curator.[8] Many of the small private museums of local societies were later absorbed by municipal governments after the Parliament enacted legislation to enable local councils to found museums.[9] However, the spread of local museums during the first half of the

century provided some limited funds for the acquisition of specimens, their preparation, and their care. The societies also rein-
forced the increasing taste for natural history, even though this
interest was often not of a very sophisticated level. In the fiftieth
report of the Leeds Philosophical and Literary Society, the Council
noted that:

It was doubtless anticipated by the originators of the Society, in conformity
with the practice of similar institutions elsewhere, that much of its time
would be occupied with papers on the Natural History of the neighborhood,
and that it would be principally occupied with the investigation of local
Geology, and of the fauna and flora of the district. Many circumstances have
contributed to the non fulfillment of this expectation, and to the attention of
the Society being directed to the other objects. Experience has shown that
the meetings of the Society, instead of being composed mainly of the students
of science, consist of a miscellaneous audience; whom papers of this kind,
being necessarily occupied with technical details, are little calculated to
interest while the establishment of Societies and of publications especially
devoted to each particular branch of Natural History of Philosophy has
attracted to them in preference, all communications connected with their
particular study. The result has been to leave such Societies as ours rather the
popularization and dissemination of knowledge than its local investigation.[10]

Edward Forbes (1815–1854) probably overstated the case in his
discussion on the educational use of museums, when he dismissed
the majority of county museums as "little better than raree-
shows",[11] however, provincial museums, with a few exceptions,
generally lacked the resources or interest for sustained research.
Their importance was more in popularizing and providing some
limited support.

Of greater importance than the establishment of provincial
museums, was the impact on the study of birds occasioned by
changes in the printing industry. In the first three decades of the
nineteenth century the printing industry was revolutionized by
the development of making paper by means of an endless web on
a machine, the application of steam to printing presses, the use of
stereotype plates instead of movable type, and cloth covers.[12]

In addition, and of special importance for ornithology, was the invention and use of lithographic techniques. William Swainson introduced their use for the depiction of birds in his *Zoological Illustrations* (1820–1823) and assured them a place in natural history thereafter.[13] The use of lithography was significant for several reasons. It allowed for a more lifelike depiction of plumage, and it generally increased the accuracy of illustrations because naturalists could draw directly onto the stone instead of relying on engravers, who often had little appreciation for scientific illustration. Also, lithographs were considerably less expensive to produce than engravings or woodcuts.

The changes in the printing industry had effects reaching far beyond the superior nature of lithographs and their commercial value. It made possible an extensive low scientific culture, which if not contributing directly to the more esoteric realm, provided support for an ever-growing number of science writers.[14] It likewise made possible the inexpensive quality bird book. This is not, of course, to claim that deluxe editions were replaced by inexpensive treatises. The splendid artistic creations of Audubon, Gould, and Lear all belong to this period, and they were quite fashionable − and costly. The point is that quality illustrations could be printed at less cost, and that a good number were. They appeared in treatises, popular guides, and encyclopedias. They also were to be found in the flood of new journals occasioned by the changes in printing and the increased market for natural history works. These new natural history journals, made possible by the potential financial return of such a venture, had a high mortality rate, especially those that tried to be rigorously scientific. *The Zoological Journal*, the first English journal devoted exclusively to zoology, informed its readers in the introduction to the first volume (1824) that:

Original Memoirs and Monographs will take the precedence in our pages. The subjects of Zoological Classification – Comparative Anatomy – particularly Classes, Families, Genera, and Species – Animal Chemistry – Palaeontography and Nomenclature are amongst the most important.[15]

The editorial policy, although it reflected what were taken to be the serious questions of the day, was not a policy aimed at a broad popular audience. Nor was the tactic suggested by Charles Konig of the British Museum in a letter dated October 3, 1809 to James Edward Smith of the Linnean Society, much better:

The best plan, perhaps, will be to make it at first as light reading as possible, to proceed gradually *ad graviora*, & lastly, when the work is perfectly established to consult nothing but the real advancement of the science, without minding the *Vulgus*.[16]

The existence of such journals at all showed the presence of a small group of serious naturalists, but unfortunately the demise of most of these journals also exposed the limits of their audience. Those that were able to stay solvent, however, played an important role for it was in these journals that brief reports of current research could be published.[17] As was mentioned in chapter five the access to publication in the late eighteenth and early nineteenth centuries was quite limited. Books had been expensive and access to publication therefore limited. With the proliferation of journals many new possibilities were opened: information could be disseminated less expensively and more widely, individuals outside a select circle could more easily transmit their findings, and knowledge gained from the research could be communicated more rapidly.

Increased empirical data, the growth and proliferation of museums, an ever-widening audience for science writers, and the consequences of changes in the printing industry all encouraged the development and diversification of interest in natural history. Ornithology came to be referred to as a department or branch of natural history, a designation that very well conveys the recognition of its degree of specialization into an area distinct from others, but that at the same time does not recognize it as a field that had the resources to stand independently as a discipline or with separate institutions such as societies, journals, etc. Specialization within natural history had not, then, grown to the point of establishing new disciplinary institutions, however, natural history institutions were altered by its various departments. Before the nineteenth century, scientific societies and publications were easily accessible to the beginner. Increased specialization and rigor altered that state. A new generation of naturalists strove to perform or found institutions along more specialized and narrow boundaries. William Swainson, for example, campaigned rigorously in Great Britain for a reform that would upgrade British scientific institutions. He wrote to Charles Babbage, who was attempting the same task, in reference to the *Preliminary Discourse* to be published in *The Cabinet Cyclopedia* edited by Dionysius Lardner:

I enter not into individual management, my [sic] chief object being to shew that none of our Societies are conducted upon principles adequate to their proposed object and that there exists a necessity either for great reform, or for the formation of a distinct Society composed *exclusively* of men of *known* and eminent talent wherein, in short, the Elite of the Science of the country should be *alone* admitted.[18]

In part the national congresses that developed, starting with the *Deutschen Naturforscher-Versammlungen* in Leipzig in 1822, attempted to fill the need for a serious public scientific forum. The *Gesellschaft Deutscher Naturforscher und Ärtze* and the British Association for the Advancement of Science (1831), modeled on the German association, were important institutions that brought together forums on a regular basis to discuss important topics and to serve as an avenue of communication. In the absence of a central institution like the French *Muséum*, the *Gesellschaft Deutscher Naturforscher und Ärtze* and the British Association for the Advancement of Science assumed an important role. In France the *Congrès scientifique de France*, started in 1833, was looked down upon by the Parisian press as not very "professional." *Le Moniteur Universel* noted in 1841 in a report on a recent Italian scientific congress that such an,

... institution, which tends to bring together distinguished and progressive men more than scientists, strictly speaking, characterizes our age. One feels the need for generalizing and popularizing human knowledge through mutual communication among the men who cultivate it; to bring together, under the auspices of science, the bonds of union among peoples.[19]

This somewhat patronizing account was not quite fair. There was an element of popularizing that went on at the national congresses, and one finds the practical sciences represented along with the pure sciences. Outside of France, these congresses served to link individuals otherwise separate. Even in France the *Congrès scientifique de France* was an important institution in that it served as a forum for minority traditions, or positions frowned upon by the Parisian professors. One finds, for example, in the annual reports of the French congress a continuing discussion on the nature of species and on the evolution of animals, a topic allegedly dead after Cuvier's debate with Etienne Geoffroy Saint-Hilaire. There are papers on migration, bird behavior, evolution of birds, etc.; topics not well represented by the publications of the *Muséum*. At the first meeting, for example, La Fresnaye, the President of the natural history section, read a paper on the classification of the passerine birds (the perching birds), the largest group of birds and the most difficult to order. Although the topic was "mainstream" the approach was not; La Fresnaye was one of the French ornithologists who, like Verreaux, looked at the birds' *moeurs* for a key to classification rather than to comparative anatomy.[20]

The spread of national congresses, with their special sections, provided a serious forum for scientific discussion. It developed at the same time that a wider popular scientific culture was coming into being, and is indicative of how the high and low scientific cultures, which in the eighteenth century were closer in aim and personnel, diverged increasingly in the nineteenth century. The history of the Linnean Society, the most important society in England for natural history in the late eighteenth and early nineteenth centuries, reflects this same trend. Several of its members felt the need in the 1820's for more specific foci. William Kirby, a leading entomologist, wrote to William Sharp MacLeay in 1822 "nothing would tend so much to the promotion of *natural science* in general as dividing the society into Committees, one for each department of Natural History".[21] The result was the short-lived Zoology Club of the Linnean Society. The restrictions placed upon the society within the society were too strict for the Zoology Club to develop, however its members soon helped found another society that did prosper: the Zoological Society of London. Due to the popular support it received from its Zoological Garden and the promise of agronomic progress, the Zoological Society was able to build an enormous collection and to engage in highly specialized zoological research.

Natural history was clearly becoming more and more specialized between 1820 and 1850. Conditions were such that a small class of professional naturalists existed.[22] Although no specialized institution existed which could be labeled "ornithological" it is legitimate to speak of such a scientific discipline.[23] How, then, shall we characterize it? *Ornithology as a scientific discipline, as it emerged in the decades between 1820 and 1850, was characterized by an international group of recognized experts, working on a set of fruitful questions, using an accepted rigorous method, and holding a common goal. It was a discipline that rested on a large empirical base and had available to it avenues of communication.*

The ornithologists who comprised this discipline were in a sense a second generation of serious ornithologists. The first generation included Cuvier, Temminck, Vieillot, and Illiger: the first ones to profit from the flood of new data, and the ones who helped set the direction ornithology took in the first half of the nineteenth century. What characterized the second generation was an increased specialization and a desire to see higher standards prevail. Comments like the following made by Brillat-Savarin in 1826 on the nature of poultry might have been quoted by Buffon but certainly not by Charles Lucien Bonaparte:

I am a great believer of secondary causation and firmly believe that this entire group of gallinaceous birds has been created solely to endow our larders and enrich our banquets.

Essentially from the quail to the turkey-cock, wherever one encounters an individual of this numerous family, one is certain to find a light food, which is tasty and which is agreeable equally to a convalescent as to a man who enjoys the most robust health.[24]

The naturalists of the period were self-conscious of their great degree of specialization and of their more scientific standing. The Rev. Leonard Jenyns in his essay "Some Remarks on the Study of Zoology, and on the present state of Science" (1837) noted that:

Natural History has not only, like most other sciences, made great progress of late years, but it has assumed an importance, to which formerly it in vain attempted to lay claim. It is not, indeed, surprising that so long as it was restricted to collecting plants and animals as mere objects of curiosity, or judged to be of no further consequence than as it admitted of application to economic purposes, it should be either held up to contempt by the majority of thinking men, or tolerated only so far as it was studied with immediate reference to the ends just alluded to.[25]

Jenyns believed that zoology was moving towards an understanding of the natural order in nature, and that one needed to study carefully the known species, rather than just naming new ones, to further our knowledge. He stressed the value of careful monographs to document specific areas, and to those just entering the field of zoology he advised: "we would recommend to such as really desire to advance its progress ... that they restrict their chief attention to some given department, and, when practicable, to those particular groups which have been least studied".[26]

The periodical literature clearly reflects the degree to which specialization had occurred by the 1830's. We can see it in F. E. Guérin's (1799–1874) journal, *Magasin de zoologie, journal destiné a établir une correspondance entre les zoologistes de tous les pays, et a leur faciliter les moyens de publier les espèces nouvelles ou peu connues qu'ils possèdent*, which published short reports with colored illustrations of new species, and for a time was divided into different sections that could be purchased separately.[27] Guérin, like other editors of the time, had difficulty supporting such specialized ventures. By consolidation with others and finally with government subsidies he managed, however, to maintain an active publishing career. His request for government support is an interesting document for it reflects the belief in the importance of such a specialized journal while also recognizing the impracticality of such a commercial publication:

The *Magasin de zoologie* is the *only* collection of this nature which is published in France and which provides to zoologists a free means of making known their work and their discoveries. Begun in 1831, and having already been in existence for fifteen years, it forms a collection of fifteen volumes, embellished with more than 1,100 plates and contains a considerable quantity of material which in all countries is used daily by the authors of general treatises, or cited by professors in their courses, and it places all persons who are seriously occupied with zoology under the obligation of consulting this collection.

The *Magasin de zoologie* is not a *Manual* or an *Elementary Treatise*, addressed to students and to persons desiring only to acquire an idea of the natural history of animals; it has a higher and more scientific goal: that of advancing zoology by enriching this science with new facts. It has, then, for readers only the elite of *scientists*, that is to say, a public small in number, in general not well-favored by wealth, a large part of whom goes to consult it in public or society libraries, which greatly limits the number of subscribers.[28]

One can find evidence of increased specialization and higher standards of scientific rigor in numerous other places. Perhaps the best reflection is in a report entitled "Report on the Recent Progress and Present State of Ornithology" given by Hugh Strickland

(1811–1853) at the British Association for the Advancement of Science meeting of 1844. Strickland confined himself to the serious "progress" of ornithology since 1830 and passed "over such works as are devoid of *scientific* mention, as well as those mere compilations, which from their want of any new or original matter tend only to diffuse and not to advance the science".[29] His subject was so large that he felt it necessary to divide it into nine headings: general systematic works, regional avifaunas, monographs on particular groups, miscellaneous descriptions of species, illustrations, anatomy and physiology of birds (mostly the results of comparative anatomy), fossils, museums, and desiderata. A reading of this fifty-page report, which attempts to summarize the *significant* work of fifteen years of research, cannot help but impress on its reader the dramatic growth that had occurred in ornithology in less than half a generation. His list of journals that were devoted to natural history and which contained original contributions in ornithology included ten British and nineteen foreign. He listed thirty-four scientific societies ranging from Moscow to Philadelphia whose publications included ornithological details, and recorded the names of scores of museums, both public and private, that contained bird collections. The most revealing section is his desiderata of which he listed eight: (1) increased precision and uniformity in naming genera, (2) a method distinguishing clearly between species and varieties, (3) information on habits, anatomy, oology and geographical distribution, (4) exploration of as yet unknown (ornithologically) areas such as China, Madagascar, etc., (5) scrutiny of names and characters of known species, (6) method of centralizing scientific information so as to avoid nominal species, (7) more studies of general ornithology or monographs (instead of "the almost exhausted subject of European or British Ornithology",) (8) more scientific arrangement of ornithological collections, both public and private.[30] It is a very revealing list in that it mostly concentrates on refining the practice of ornithology. The main issue in classification was defined as the working out of genera and of defining species, rather than of obtaining new ones or of establishing broad outlines of a new general classification. Collections needed to be put into order, rather than expanded. There was even the recognition of the need for natural history studies, although these were seen as important mostly for the light they might shed on classification.

Strickland's review article is notable for covering a wide range of topics over a relatively long period of time. Journals such as the *Archiv für Naturgeschichte* (founded in 1835) contained yearly assessments of the "progress" of zoology, and documented the important work done in each department. It is not difficult to perceive in Strickland's report, or the reports in periodicals, a central core of internationally-recognized ornithologists: in Britain there were Gould, Macgillivray, Yarrell, Jardine, Selby, Swainson, Vigors, George Robert Gray, Strickland, Blyth; in France: La Fresnaye, Geoffroy Saint-Hilaire, Lesson; in Germany: Lichtenstein, Nitzsch, Prince Maximilian, Brehm, Naumann, Rüppell, Spix, Wagler, Kaup; Sweden had Nilsson and Sundevall; Austria had Natterer; Holland still had Temminck and later Schlegel; the United States had Audubon; and Italy had Bonaparte. All of these individuals were professional in the sense that they were either independently wealthy but expended the amount of time and effort equivalent to a gainfully-employed individual, or they were able to support themselves directly or indirectly by their work in ornithology. The majority were associated with one of the several large museums that had professional staffs. What held this disparate and widely-scattered group together was not a central institution, international society, main publication, or founding father, but rather a set of fruitful questions and a common general goal.

The central questions of ornithology radiated from classification. Documentation of the wealth of material was basic.

Buffon's *Planches enluminées* was the foundation for iconography and the splendid French bird books of the early part of the century complemented them. In 1820 Temminck, in collaboration with the French collector Baron Meiffren Laugier de Chartrouse, began a continuation of Buffon's plates in the *Nouveau recueil de planches coloriées d'oiseaux*, which ultimately depicted an additional six hundred and sixty-one species by 1839. Like Buffon's plates, Temminck's quickly assumed an important role. An 1839 report by the curator of the Zoological Society of London, for example, complained that "Great difficulty is experienced in naming birds from the want of Books. The Planches Coloriées [sic] of Temminck is quite essential and the want of this work was felt by Mr. Gould, who stated before his departure, that it would be impossible to proceed without it."[31] Plates, however, never replaced specimens in importance, particularly type-specimens, and major ornithologists like Temminck, Bonaparte, and Gould often found it necessary to travel extensively among the public and private collections of Europe. For everyday practice, however, books had to suffice. Fortunately, the improvements in printing, the public's taste for luxury natural history books, and the quality of ornithological investigation produced a happy confluence of forces and resulted in a set of luxury bird books that satisfied both the scientific desire for documentation, and the public's fancy. Most well known among these splendid creations were those done by Audubon and Gould.

John James Audubon's (1785–1851) artistic rendering of North American birds, often employing dramatic effects, was quite a sensation in Europe. Unable to raise enough interest in the United States for his drawings, Audubon traveled to Great Britain where they were engraved and published and where his work was appreciated. His most famous collection was *The Birds of America* (1827–1838), which contains four hundred and thirty-five plates with one thousand and sixty-five figures of four hundred and eighty-nine species, and is one of the most splendid works in bird art. They were published in double elephant folio in order to depict the birds in their natural size. *The Birds of America* was also notable because unlike most of the grand bird art books the drawings were done from living or freshly-killed specimens.[32] Audubon was in many ways outside the mainstream of ornithological research for he was untutored in systematics and nomenclature. He was, however, a great artist and an attentive field observer. His five-volume *Ornithological Biography* published between 1831–1839 was edited, fortunately, by William Macgillivray (1796–1852) whom Audubon cited in the preface to the first volume as "completing the scientific details and smoothing down the asperities of my Ornithological biographies".[33] It was not, however, these details that were most important. Frédéric Cuvier (1773–1838), in a very perceptive review of the *Ornithological Biography*, wrote:

Monsieur Audubon is not, however, a naturalist, he is a skillful painter and an intelligent observer. Perhaps it is exactly because he is a stranger to the study of nature that he was led to create an original work of natural history that no professional naturalist would probably have the idea of attempting. For the direction given to science, and which more or less all those who are concerned today with the study of natural beings follow, does not lead to the researches to which Monsieur Audubon devotes himself.[34]

Audubon documented the behavior and life histories of birds in a way that the majority of the ornithologists of his day ignored. In his review of Temminck's plates Frédéric Cuvier, who along with Verreaux and La Fresnaye worked in this minor tradition, lamented the direction specialization had taken "natural history":

Up to now Monsieur Temminck is still only at the introduction of the natural history of birds; in order to give us knowledge of them, it would remain for him to give us the natural history, properly speaking, of them. For without the knowledge of living birds — active, realizing the designs of nature and their destiny on earth, fulfilling, in a word, the role that they have to play in the general economy of nature — ornithology is only an incomplete science, which leads us only to the entrance of the structure; we admire the outside, and we are left entirely ignorant of all the riches that it contains.

Unfortunately, today that is the character, not only of ornithology, but also of all the natural sciences. The only goal naturalists have is to collect species and to compare them according to their degree of resemblance. To study the actions, to discern the character, distinguish the instincts, study the extent of intelligence; to appreciate the relations of these moral phenomena, identify their reciprocal influence, and to infer the object of animals and the designs of Providence when it created them; are ideas which have almost entirely become foreign to science. No school professes them, and they assume no part in intellectual life. It is not, then, surprising that Monsieur Temminck keeps almost silent on the natural history of the species he publishes. Working in his cabinet, he can consult only his collections and the naturalists who have written on birds.[35]

The bird books of John Gould have similarities to both Temminck's and Audubon's. They are large folio volumes that allow for life-size drawings and were important for iconographic documentation.[36] Like Audubon, Gould's work was a commercial venture. He published his own books by organizing the activities of collectors, lithographers, and printers.[37] As for the plates, Gould made sketches that he later worked into watercolors, which in turn were put onto stone by one of a number of successive and highly-skilled illustrators, most notably his wife Elizabeth, and then after her death, Henry Richter (b. 1821). Some of the early works were assisted by Edward Lear, famous for his *Illustrations of the Family of Psittacidae, or Parrots* (1832). Like Audubon, Gould also did some field work himself, most notably in Australia where he recorded three hundred new species. However, unlike Audubon, Gould was more sensitive to the questions that occupied naturalists and was more a part of the discipline of ornithology. He had begun his career at twenty-three as a taxidermist for the Zoological Society and quickly rose to the superintendentship of its ornithological collection. Consequently, he was ideally placed to become acquainted with systematics. He also, in connection with his position, had the opportunity to travel to examine some of the major collections on the continent. By 1836 he was competent enough that he was requested by Charles Darwin, recently back from the explorations of *H. M. S. Beagle*, to work on the bird collection for publication.[38] Gould's own publications then, like Temminck's, were important for their contribution to systematics as well as iconography.

The documentation supplied by Audubon, Gould, Temminck, and others was largely of external morphology. Information on the natural history of individual species, although recognized as interesting and useful for some future complete natural history of birds and included when possible, was not widely pursued. Similarly, although information was accumulating and being collected, the study of bird distribution was secondary. What ornithologists primarily focused their attention upon were the technical problems of classification. There were four principal issues: distinguishing between varieties and species, establishing genera, constructing a natural system, and standardizing nomenclature. The first of these rose to prominence as a side effect of the growth of detailed information on birds. Edward Blyth's (1810–1873) articles published in *The Magazine of Natural History* (1835 and 1836) show how sophisticated the discussion had become. Blyth distinguished seasonal, age, and sexual variations from simple variation of color and size, acquired variation, and finally true varieties.[39] Although his position was very conservative in stressing the integrity of the species and practically reducing what we would call a variety to a null set, his discussion reflected a vast quantity of empirical information from which such distinctions could be made. It also reflected the static approach of most of ornithology before Darwin, and the tendency to proliferate species, i.e., split groups into even smaller units and consider them as true species. This approach was carried perhaps to its limit by Ludwig Brehm. Unlike Buffon who believed that climate and environment could produce varieties, Brehm believed that species were immutable, therefore any geographic variety was a distinct species. Herman Schlegel, Temminck's successor at Leyden, invented a trinomial nomenclature to simplify naming them.

More important, and more widely discussed than the splitting of species, was the break up of the unmanageable Linnean genera. The detailed information contained in monographs and the ever-growing empirical base made the splitting of genera a practical and intellectual necessity. In 1820 Latham wrote to James Edward Smith that he was shocked by Temminck's use of slightly over two hundred genera, almost twice what he had been forced to use. By 1844, George Robert Gray (1808–1872), in his *The Genera of Birds* described eight hundred and fifteen, in an attempt to bring order to the over two thousand, four hundred genera names compiled from other authors.

The quantity of work done on the lower taxonomic levels, *i.e.*, the monographs on individual genera and families, the splitting of genera, and the description of hundreds of new species, brought into stronger relief the need for an overall natural system of bird classification. The *Règne animal* of Cuvier exerted great influence, as had Linnaeus's earlier *Systemae Naturae*, because of its completeness and its overall structure. However, the arbitrary and artificial aspects of his detailed arrangement of the birds were evident. One response as we have already discussed, was to pursue the anatomical investigation to finer resolution and uncover a uniform basis for avian classification. Blainville, Nitzsch and Carl Jacob Sundevall (1801–1875), pursued that line of research, which later in the century bore fruit. Another response was to supplement comparative anatomy with knowledge of the birds' *moeurs*. La Fresnaye and Verreaux took that approach. However, the lack of available data inhibited its development.

What appeared at mid-century to many as a reasonable and potentially fruitful program for uncovering the system in nature was the study of affinities, which basically was an attempt to incorporate all that was known about birds into a single system. The appeal of the approach is indicated by the enormous variety of individual systems and the wide range of principles employed. Hugh Strickland, whose report has just been discussed, attempted to map affinities in an inductive fashion using physical characteristics reminiscent of Linnaeus's sketches of a natural system for botany.[40] Strickland wanted to build the system from the base upwards; that is, to start with individual species and compare them with those closest in similarity of essential features (in terms of comparative anatomy). "The natural system may, perhaps", he wrote, "be most truly compared to an irregularly branching tree, or rather to an assemblage of detached trees and shrubs of various sizes and modes of growth".[41] In what, from an historical perspective, may appear as an unintentional and amusing throwback to the days of Pallas, Strickland added:

If this illustration should prove to be a just one, the order of affinities might be shown in museums in a pleasing manner by constructing an artificial tree, whose ramifications should correspond with those of any given family of birds, and by then placing on its branches a stuffed specimen of each genus in their true order.[42]

Strickland was not proposing a genetic relationship, and he employed metaphors other than organic ones. For example, he wrote that:

All that we can say at present is that ramifications of affinities exist; but whether they are so simple as to admit of being correctly depicted on a plane surface, or whether, as is more probable, they assume the form of an irregular solid, it is premature to decide. They may even be of so complicated a nature that they cannot be correctly expressed by terms of space, but are like those algebraical formulae which are beyond the powers of the geometrician to depict.[43]

Strickland's approach had a lot to recommend it, and it appeared to be appropriate given the situation. It was related to the tradition of the Blumenbach school of using the total *habitus*, which can be seen in Illiger's work, as well as in the independent but related work of some of the Paris anatomists such as Isidore Geoffroy Saint-Hilaire.[44]

In contrast to the cautious inductive approach of Strickland, certain authors who were looking for affinities thought that in so doing they had uncovered a simple and universal regularity in nature. William Sharp Macleay (1792–1865) in his *Horae Entomologicae; or Essays on the Annulose Animals* (1819) argued that there is an order among living beings that can be thought of as circular. By tracing affinities, it appeared that the relationships within a group return to the starting point. One encountered thereby a pleasing symmetrical relationship. In the 1820's Macleay's ideas were extended to birds by Nicholas Aylward Vigors (1787–1840) in a set of papers he delivered to the Linnean society. In these papers, and in a set published in the *Zoological Journal*, Vigors sketched a general system for birds that was based upon Macleay's formula of telescoped circles each of five units, hence the general name of the quinary system. Part of the appeal of the system was that it not only organized the material but was also weakly predictive. In following the ramifications from five orders, each composed of five families, containing five genera, etc. (or inventing finer subdivisions), one encountered blank spaces unfilled by any known group, presumably yet to be discovered.

The quinary system was attractive for it seemed to work for the animal kingdom as a whole, it was aesthetically pleasing, and it was reconcilable with the Linnean and Natural Theology traditions.[45] Moreover, in the manner in which it was elaborated, it seemed to be able to absorb, not only information from comparative anatomy, but even the fragmentary information on distribution, behavior, etc. William Swainson in a number of publications tried to develop the general outlines of such a system. Swainson divided his quinary circle into three groups: typical, subtypical, and aberrant, this last in turn compounded of those groups of the same rank as typical and subtypical. The members of a typical group:

> . . . are the most perfectly organized: that is to say are endowed with the greatest number of perfections, and capable of performing to the greatest extent, the functions which peculiarly characterise their respective circles. This is universal in all typical groups; but there is a marked difference between the types of a typical circle, and the types of an aberrant one. In the first we find a combination of properties concentrated, as it were, in certain individuals, without anyone of these preponderating, in a remarkable degree, over the others; whereas in the second it is quite the reverse: in these last, one faculty is developed in the highest degree, as if to compensate for the total absence, or very slight development, of others.[46]

A high degree of development of a single aspect characterizes the aberrant circle as a whole, and the three subcircles of the aberrant group express this development differently. The three subgroups are aquatic (live in water, have enormous bulk, large heads, no or slightly developed feet, and are carnivorous), suctorial (imbibe food by suction, are of small size, defenseless, and defective in organs of mastication), and rasorial (are of large size, have developed feet, tails, and head appendages, and a superior degree of intelligence and docility). Subtypical forms are intermediate between typical and aberrant. They are characterized by being,

> . . . the most powerfully armed, either for inflicting injury on their own class, for exciting terror, producing injury, or creating annoyance to man. Their dispositions are often sanguinary; since the forms most conspicuous among them live by rapine, and subsist on the blood of other animals. They are, in short, symbolically the type of *evil*.[47]

Swainson's overall characterization admitted of a great deal of flexibility, for he held that each individual animal partakes of nine telescoping types, and is related to other animals in nature outside its circle by a complicated set of analogies based on internal and external physical characteristics, function, and habit. He applied his neo-Baroque version of the quinary system to the classification of birds in his two-volume treatise *On the Natural History and Classification of Birds* (1836).

The British naturalists who championed the quinary system tried to argue that it was grounded in observation and deduced from the facts. They were accused, however, by men such as Strickland, of having constructed an *a priori* theory that was found "when tested by reason, to be improbable, and when by observation, to be untrue".[48] In spite of such criticisms, the quinary system enjoyed a considerable vogue.

The search for a fundamental regularity that would yield a key to the natural system of classification was also a major concern of certain German naturalists. In addition to the systematics done by Illiger and Wagler, there was an attempt, influenced by the *Naturphilosophen* inspired by Friedrich Wilhelm Joseph Schelling's (1775–1854) writings on nature, to devise a system from first principles. All of these systems, of which the most well known were done by Lorenz Oken (1779–1851), Reichenbach and Kaup, can be characterized as speculative, hierarchical, and focused on anatomical features, starting with man as the ideal animal and using his characters to define general groups. Kaup, for example, used the number five, like Swainson, as a basic unit, but defined the five components by the five senses of man.[49] There was no attempt by these naturalists to justify their system as "deduced from the facts"; quite the contrary, the facts were "deduced" from the first principles.

The lack of agreement on the foundations of a natural system, even on the method of establishing such foundations, should not cause us to overlook the fact that it was such a search that occupied a significant portion of naturalists throughout the ornithological world. The situation was not quite as chaotic as one might suppose. The broad outlines of a system of birds as sketched by Cuvier and Temminck were recognizable in most "new" systems and the majority of ornithologists were eclectic. The genera of birds as compiled and established by Gray soon achieved a level of authority.[50] Temminck had updated Latham's *Synopsis* in the second edition of the *Manuel* by adding an index, and Charles Lucien Bonaparte was traveling to every major collection in Europe in order to produce a more up-to-date species list.

The greatest success classification achieved in the first half of the nineteenth century was in the standardization of nomenclature. Because such a standardization was a recognizably needed convention, rather than an empirical or theoretical issue, it was less hotly debated than the foundations for a natural system. As the preface to the "Report of a Committee appointed 'to consider of the rules by which the Nomenclature of Zoology may be established on a uniform and permanent basis'" stated:

All persons who are conversant with the present state of Zoology must be aware of the great detriment which the science sustains from the vagueness and uncertainty of its nomenclature. We do not here refer to those diversities of language which arise from the various methods of classification adopted by different authors, and which are unavoidable in the present state of our knowledge. So long as naturalists differ in the views which they are disposed to take of the natural affinities of animals there will always be diversities of classification, and the only way to arrive at the true system of nature is to allow perfect liberty to systematists in this respect. But the evil complained of is of a different character. It consists in this, that when naturalists *are* agreed as to the characters and limits of an individual group or species, they still disagree in the appellations by which they distinguish it. A genus is often designated by three or four, and a species by twice that number of precisely equivalent synonyms; and in the absence of any rule on the subject, the naturalist is wholly at a loss what nomenclature to adopt. The consequence is, that the so-called commenwealth of science is becoming daily divided into independent states, kept asunder by diversities of language as well as by geographical limits. If an English zoologist, for example, visits the museums and converses with the professors of France, he finds that their *scientific* language is almost as foreign to him as their *vernacular*. Almost every specimen which he examines is labeled by a title which is unknown to him, and he feels that nothing short of a continued residence in that country can make him conversant with her science. If he proceeds thence to Germany or Russia, he is again at a loss: bewildered everywhere amidst the confusion of nomenclature, he returns in despair to his own country and to the museums and books to which he is accustomed.[51]

It was clearly in everyone's interest to standardize nomenclature, and the international character of the emerging zoological disciplines made it a practical necessity. The British Association for the Advancement of Science hoped that a set of rules proposed by one of its committees would be:

... invested with an authority which no individual zoologist, however eminent, could confer on them. The world of science is no longer a monarchy, obedient to the ordinances, however, just, of an Aristotle or a Linnaeus. She has now assumed the form of a republic, and although this revolution may have increased the vigour and zeal of her followers, yet it has destroyed much of her former order and regularity of government. The latter can only be restored by framing such laws as shall be based in reason and sanctioned by the approval of men of science.[52]

The report of the committee was successful; it established the law of priority, utilized Linnaeus's binomial nomenclature, and designated the twelfth edition of *Systema Naturae* as the starting point of reference. It also established rules and suggestions for orthography and made recommendations for the general improvement of nomenclature. The report was in the best Linnean tradition: clear, practical, and simple. It is not surprising that it quickly was accepted and utilized.

The reform of nomenclature gave zoology, and ornithology in particular, a common language, which not only facilitated communication and removed much unnecessary confusion, but also was a major step in helping ornithologists towards the common goal they shared: a complete catalogue of the birds of the world. Two generations earlier, Buffon and Brisson, each in his own fashion, had envisioned such a compilation. The ornithologists of

the middle of the nineteenth century could see one in the not
too distant future. It would be clearly in the style of Brisson, that
is, a list and characterization of the species and genera of birds.
What would not be in the catalogue was revealing. A full discussion
of each bird's natural history, although desired, was still something
for the distant future. Although anatomical considerations were
important for most classification schemes, anatomy and physiology
per se would for the most part not be visible in a bird catalogue.
They were separate disciplines and the ornithic material employed
in them was primarily for elucidating anatomical and physiological
problems. Nor would there likely be any lengthy discussion of the
relationship of birds to other animal groups, for that was in the
realm of general zoology and systematics, in which there was
little common agreement.

That this common goal, a general catalogue of birds, was
thought not far off can be seen in the beginnings of the great
catalogues of the mid-century. Temminck had attempted a list
of species in the index to the second edition of his *Manuel* (1820–
1840), which supplemented Latham's *Synopsis*. In 1840 George
Robert Gray published *A List of the Genera of Birds*, which had
grown out of the work done for the rearrangement of the ornithology collection of the British Museum. Gray attempted to list all
the legitimate genera names to date and to clarify synonyms. He
expanded his list a few years later into *The Genera of Birds;
Comprising Their Generic Characters, a Notice of the Habits of
Each Genus, and an Extensive List of Species, Referred to Their
Several Genera* (1844–1849), which became one of the standard
reference tools throughout Europe.

The most detailed and ambitious of all the bird catalogues was
the incomplete *Conspectus generum avium* (1850) prepared by
Charles Lucien Bonaparte (1803–1857). In it he attempted to
include all known species, which by then were thought to number
over seven thousand forms. The task of compiling and comparing

Illustration 10. "Bonaparte" lithograph (1849) by J. H. Maguire. (author's collection).

descriptions, examining specimens, and establishing synonyms was Herculean and took a Bonaparte to tackle it. He was well qualified.[53] Ornithology had been his passion from youth, and he was fortunate to have the means to pursue it fully. As a young man, he visited the United States and soon set out to complete Wilson's *American Ornithology*, which he did admirably, not by field research but by museum study. When he returned to Europe, he visited major collections, public and private, and soon was regarded as one of the leading experts in ornithology.[54] He fully utilized the magnificent collections in Leyden, Berlin, and London, and ultimately settled in Paris where he worked at the *Muséum*. Bonaparte was aided in his research, not only by his ability to travel to major collections, but also by a series of museum catalogues that started to appear towards the middle of the century. In 1844 the first section of a *List of the Specimens of Birds in the Collection of the British Museum* appeared, and Lichtenstein in Berlin produced a series of catalogues, the most well known, the *Nomenclator Avium Musei Zoologici Berlinensis* (1854) described a collection of over four thousand species. These catalogues were useful guides to what existed in different collections and more importantly, since they usually identified the source of the specimen, they served as indices to type-specimens. The value of these catalogues was well described by Richard Owen (1804–1892), later first director of the British Museum (Natural History), in his testimony before the House Select Committee where he responded to a question on the importance of collection catalogues. "I consider", he said, "that in a national collection of natural history it is quite essential; that such a catalogue constitutes, in fact, the soul of the collection."[55]

Bonaparte's death deprived the scientific world of a completed *Conspectus generum avium*. What he had accomplished was appreciated as witnessed by the set of letters addressed to the government from every major museum and from the recognized leaders in the field that Verreaux assembled in support of a project to finish Bonaparte's bird list.[56] The *Conspectus generum avium* remained, however, uncompleted. Nonetheless it stood as an indication that a complete catalogue was in sight; a goal that in Paris could only dimly be imagined less than a century earlier by Buffon or Brisson.

A lot had changed, of course, from the time of Réaumur, Brisson, and Buffon. As we have seen in the previous chapters, the empirical base of ornithology had grown prodigiously, and natural history collections had been transformed from curiosity-cabinets to working museums. Literature on birds was equally transformed; there was more of it, and it tended to be more specialized. In part, the specialization was the natural consequence of a number of technical and social factors. European economic expansion and changes in the printing industry and public taste created greater possibilities for professional careers, for publications of narrow focus, and for communication among ornithologists. After 1830 there existed an international community of recognized experts who were in correspondence and who shared a common rigorous method for studying what they perceived to be an interesting set of questions. These questions emanated from classification, for it was the classification of an ever-swelling quantity of bird specimens that was most pressing. It was this concern with classification and the object of a complete catalogue of birds that, in essence, defined ornithology; that is, demarcated it from the study of other living beings. Ornithology, as a scientific discipline, had come into existence, and like many of the other scientific disciplines of the nineteenth century, once emerged, developed rapidly: journals, societies, catalogues, and an increase in publications followed rather quickly. An ornithologist of the second half of the nineteenth century, if asked what was known about birds, would have responded by saying that a *great* deal was known about them. In so saying he would have meant something quite different from a gentleman responding in the same manner to the same question a century earlier. The difference between their two answers reflects the difference in the perception of birds occasioned by the emergence of the scientific discipline of ornithology.

CHAPTER VIII

THE SIGNIFICANCE OF THE EMERGENCE OF ORNITHOLOGY AS A SCIENTIFIC DISCIPLINE

To better appreciate the change that took place in less than a century in the study of birds we might contrast the two works that commenced this study with the one that concluded it. The differences between ornithology in the second half of the eighteenth century and ornithology in the middle of the nineteenth century are, indeed, well exemplified by comparing Brisson's *Ornithologie* (1760) and Buffon's *Histoire naturelle des oiseaux* (1770–1783) with Bonaparte's *Conspectus generum avium* (1850).

Both Brisson and Buffon were acutely aware of their limitations, and they recognized that they were merely providing a new beginning point for the study of birds. Their empirical base was narrow: the Reaumur collection, other Paris cabinets of note, a small number of publications by previous authors, and a limited scientific correspondence. Nonetheless, Brisson and Buffon were able to nearly quadruple the five hundred species and varieties described by Ray. They also provided their readers with lengthy descriptive passages and scientifically accurate engraved plates. The two works were intended by their authors for a general natural history audience, and they could be appreciated by any moderately well-educated layman who had an interest in nature. Buffon's volumes had wide appeal, for they were cast in elegant prose and, like Brisson's ornithology, they were an elegant publication, printed on hand-made paper and bound in gold-tooled leather.

Charles Lucien Bonaparte's *Conspectus generum avium*, two simple octavo, cloth-bound books printed on mass-produced paper and lacking any stylistic or aesthetic grace, stands in stark contrast with the sets of French rococo quarto tomes by Brisson and Buffon. The difference reflects the shift in ornithology from the salon to the study. Bonaparte's work also reflects that ornithology had gone from a local individual enterprise to an international and disciplinary one, the magnitude of which was considerably larger than before. Although Bonaparte had extensively utilized Parisian collections (by his time greatly expanded), he examined, in addition, "the museums and forests of Europe and America",[1] especially the magnificent Leyden collection with its many unnamed exotics. He was able thereby, in turn, nearly to quadruple the number of species described by Brisson and Buffon. Unlike his predecessors' work, Bonaparte wrote for an educated specialist. To appreciate the *Conspectus* one needed the training of an avian taxonomist. Its eight hundred pages are wholly given to meticulous lists of species, synonyms or synopses, and indications of distribution. Its organization draws on the sophisticated systematics of the time, and the individual entries display the vast literature on birds available by the mid-nineteenth century. Far from considering himself a pioneer in ornithology, Bonaparte aimed at providing a nearly complete list of birds. His *Conspectus* is incomplete because he died in 1857 before he finished the project. Even incomplete it was recognized by his peers as a major accomplishment, and they appreciated it as an important step towards a comprehensive catalogue of birds, one of the several central goals that guided the discipline of ornithology.

Ornithology had indeed changed in the century after the appearance of Brisson's *Ornithologie*. It had gone from a subject typified by collection-catalogue natural history and encyclopedias that were written for a general audience, to a subject that reflected the existence of a scientific discipline in which the methods of research were rigorous and narrowly prescribed, the topics limited to a few agreed upon significant ones, and the audience a highly critical and specialized group of trained individuals. However, the story of what took place in ornithology in the late eighteenth and early nineteenth centuries has greater significance than merely clarifying how the study of birds went from a very limited activity to become a major scientific discipline. Because ornithology was among the first departments of natural history to emerge as an independent discipline, and because it was central in the theoretical debates, institutional developments, and popular appeal of natural history, this monograph is of considerable historical interest as a case study in the history of natural history in the late eighteenth and early nineteenth centuries. As such, it graphically demonstrates the inadequacy of the two current interpretations of that history. In the Introduction those two views were described. One depicts natural history as having given way to biology, *i.e.*, it advances the notion that an epistemological and methodological shift allegedly occurred within the life sciences with the result that instead of merely cataloguing nature, scientists began to consider "life" itself as a subject for research. Michel Foucault strikingly states this position in the following manner:

Historians want to write histories of biology in the eighteenth century; but they do not realize that biology did not exist then, and that the pattern of knowledge that has been familiar to us for a hundred and fifty years is not valid for a previous period. And that, if biology was unknown, there was a very simple reason for it: that life itself did not exist. All that existed was living beings, which were viewed through a grid of knowledge constituted by *natural history*.[2]

In Foucault's opinion the general epistemological position which had characterized the "Classical period" broke down at the end of the eighteenth century and was replaced by a different intellectual order.

The sciences always carry within themselves the project, however remote it may be, of an exhaustive ordering of the world; they are always directed, too, towards the discovery of simple elements and their progressive combination; and at their centre they form a table on which knowledge is displayed in a system contemporary with itself.[3]

"The last years of the eighteenth century are broken by a dis-
continuity"[4] according to Foucault, and in place of order, the
basic principles of conceptualization became function and succession. In the study of plants and animals this intellectual shift took
the form of a breakdown of natural history and the emergence of
a dynamic biology. Foucault's analysis is intellectually stimulating
and presents an interesting and new periodization of intellectual
history. He also groups together sets of savants in a novel and
suggestive fashion. However, his characterization bears little
resemblance to the historical record.

Natural history of the eighteenth century was not merely a
facet of a philosophic movement or a manifestation of a unified
worldview as much as it was an enterprise encompassing different
research traditions with different goals, methods, and perspectives.[5]
At least four distinct research traditions existed in natural history
during the second half of the eighteenth century. The one most
often used to characterize the entire enterprise concentrated on
nomenclature and systematics. Linnaeus and Brisson are good
examples of this approach. Buffon represents an opposing tradition; the attempt to construct a complete natural history of each
species. Buffon, further, held that species had to be studied
with an awareness of their temporal dimension. Only by such a
broad approach, he believed, could one understand contemporary
forms and their distribution, as well as provide the foundations
for a science that could discern the general laws regulating living
forms.

There existed two other major traditions in eighteenth-century
natural history, neither of which has been treated very extensively
in this monograph but which need to be briefly described here.
One was the comparative study of animal anatomy which later
became the independent discipline of comparative anatomy.[6]
Naturalists such as Louis-Jean-Marie Daubenton (1716–1800)
believed that comparative morphological analysis would reveal

the plan and workings of the animal body,[7] and although he did little work on birds, his protégé, Felix Vicq-d'Azyr (1748–1793), published extensively on the comparative study of avian anatomy. The other tradition of eighteenth-century natural history, which like comparative anatomy later became an independent discipline, was physiology. Naturalists such as Lazzaro Spallanzani (1729–1799) pioneered the experimental and comparative investigation of basic vital functions common to all organisms with the goal of discovering the unifying principle of living beings. In the process Spallanzani occasionally studied various aspects of avian physiology; he did so, however, for its comparative value rather than as a contribution to ornithology.

Neither comparative anatomy nor physiology has figured very centrally in this monograph on the emergence of ornithology as a scientific discipline. Instead, the two traditions of eighteenth-century natural history — complete natural histories of individual species and systematics — that ramified into the specialized discipline of ornithology have been followed. Avian anatomy and physiology were segments of two other traditions that diverged into separate disciplines themselves. Ornithologists, of course, utilized the research in these separate areas, especially from comparative anatomy, which came to be considered important for systematics. The comparative anatomists and physiologists, however, studied birds for different ends than did ornithologists. Morphologists and physiologists sought general laws regulating form and function, whereas ornithologists sought to classify and describe a particular group of animals.

To suggest that biology replaced natural history in the late eighteenth century is to arrange the contemporaneous research traditions of the eighteenth century into a chronological order. Such an arrangement has no basis in the historical record. It overlooks or denigrates the work of careful experimental physiologists of the eighteenth century, such as Spallanzani, and it ignores the vast field of nineteenth-century natural history – the work of ornithologists, entomologists, ichthyologists, etc. Foucault may be correct in suggesting that certain individuals made epistemological shifts that influenced their perception of nature, methodology, and choice of scientific questions. But as a general characterization, the notion that natural history was transformed into biology simply won't stand. Less extreme interpretations, such as the following one made by William Coleman in his history of nineteenth-century biology, need to be reevaluated as well: "Natural history remained a prosperous occupation throughout the nineteenth century . . . but the ascent of plant and animal physiology was more dramatic and it offered all the appeal of a new and potentially fundamental science."[8] Such an assertion has never been documented. From this case study of the emergence of ornithology it is evident that natural history in the nineteenth century was a highly active field, with a large institutional base, with links to commerce and colonialization, with a broad popular appeal, and with interesting and significant theoretical questions.

The failure to appreciate nineteenth-century natural history has many causes. Some historians have been taken in by the polemics of nineteenth-century advocates of biology. Others undervalue natural history because of contemporary biases. Certain historians have tried to impose a specific philosophy of history on the facts and have been led thereby to ignore a large body of them. Few have appreciated the overall historical importance of the development of natural history, for at present very little serious work has been directed towards it. Those who have studied the history of natural history tend to stress the second characterization of the transformation of natural history in the late eighteenth and early nineteenth centuries: an alleged shift from a static catalogue of nature to a history of nature. This case study of the emergence of ornithology as a scientific discipline helps to illustrate that this second interpretation is equally faulty. Buffon, Kaup, and

Geoffroy Saint-Hilaire all accepted the notion that species of birds have changed and that a full understanding of the relationships among present forms had to rest on an understanding of the past. By contrast, Brisson, Latham, and Swainson saw no need for a historical approach and developed adequate systems without it. One can list major figures, then, both in the late eighteenth and early nineteenth centuries who advocated static or dynamic views. After the publication of *The Origin of Species* the situation was fundamentally altered in many countries. But that was after 1859, well past the period under discussion here. In ornithology of the late eighteenth century one doesn't perceive a general intellectual shift. The work done then, and in the early nineteenth century, was a continuation of a line of work that started with Brisson and Buffon. There were, to be sure, conflicting philosophical positions, different theoretical biases, opposed interpretations of data and of what constituted data, and different criteria for nomenclature. But overall there was the growth and development of a scientific discipline that had at its core goals that had been laid down in the second half of the eighteenth century. The breadth of individual research narrowed, and the rigor of method increased. Judging, then, by the history of ornithology, the transformation of natural history did not result in its dissolution and replacement by an entirely new subject, but rather in its extension, specialization, and growth into separate scientific disciplines. Indeed, it was the attention paid to traditional problems of specific groups, such as birds, that helped lead to the emergence of numerous separate disciplines, for the amount of material and the care with which it was studied necessitated specialization. The events that resulted in the emergence of ornithology as a scientific discipline had little to do with epistemological shifts. It was the enormous expansion of the study of birds, made possible by the complex set of factors described in the previous chapters, that led to the specialized discipline of ornithology. Far from *das Ende der Naturgeschichte* there was a flowering of natural history in the late eighteenth and early nineteenth centuries. To understand that history one needs to go beyond simplified generalizations about worldviews. This case study suggests some issues that when investigated might help us understand that flourishing of natural history. To name but a few: the history of the emergence of other disciplines such as entomology and ichthyology and the interaction of these new disciplines is necessary to establish how general was the pattern of ornithology's development and to discern similarities that may not be obvious from a look at the history of one particular discipline. The enormous fashion of popular natural history and its relationship to more scholarly natural history has barely been investigated.[9] The relationship of natural history to applied science, for example, economic botany, might disclose some of the roots of the increase in government patronage that natural history came to enjoy. A closer look at the intellectual specialization and the professionalization in general during this period might, in addition to providing some insight into the history of natural history, take the story out of its parochial domain and help to relate it to general history.

This monograph contributes to a better understanding of the transformation of natural history in the late eighteenth and early nineteenth century by providing a case study of a central event in that transformation and it suggests questions and topics that need further scrutiny. The history of the emergence of ornithology as a scientific discipline bears also on several other general topics in the history of science. One such issue, which has been discussed at length in recent years, is the professionalization of science in the nineteenth century.[10] Although many historians, sociologists, and scientists agree that a salient feature of nineteenth-century science was its newly acquired professionalism, the subject is one of which at present there is little understanding.

There is no general consensus concerning its causes, development, or even on the definitions of the basic terms with which to discuss it. Some very careful studies and distinctions have nonetheless been made. Nathan Reingold, for example, divides the pre-1900 scientific community in the United States into cultivators, practitioners, and researchers.[11] His characterization avoids the anachronisms of classifications based on analyses of modern professions, yet unpatronizingly allows for distinctions among individuals who were to a greater or lesser extent involved in science. He also cautions against oversimplification of the story.[12] But oversimplification of such a complex problem is difficult to avoid. As Reingold points out: "Historians and scientists have at least one thing in common — both like to handle problems by reducing the number of variables involved."[13] Historians, for example, have been too hasty in generalizing from studies of single countries. For instance, Maurice Crosland, after describing the developments of a professional scientific career in France, suggests that British professionalism in science lagged behind the French by fifty years.[14] Crosland maintains that the French model was significant for British Science, nonetheless, for it was one that could be later adopted and adapted by the British in their own way. Only a comparative analysis, however, can justify such an inference. A similar sort of problem exists with Joseph Ben-David's widely cited *The Scientists' Role in Society. A Comparative Study* which draws from studies that have been done on individual countries and which arranges the cases sequentially but with very little real comparative analysis.[15] The reader is given the impression by the organization of the material that he is following a chronological development of the professionalization of science. However, a detailed study of even one country usually dredges up a rather murky picture that obscures clear outlines of development.[16] This is especially the case if the study concerns the period before 1850 when the very term "professional" is highly problematic to apply in science for most countries. In Great Britain, for example, the first systematic attempt to count the members of the professions took place in the Census of 1841 where the only professions listed were church, law, and medicine. Under another heading, "Other Educated Persons", a number of other professions, recognizable then or later, appeared, including six ornithologists.[17] Given that the census listed 63,000 professionals in Britain, the number of ornithologists recognized by the general public was pitifully small, hardly enough to constitute a "profession".

We know from this study that even if a profession of ornithology did not exist, considerable ornithology of a high caliber was done in Britain in the second half of the nineteenth century. Perhaps there is a useful distinction between a profession and a discipline to be drawn from the history of ornithology. This monograph has stressed the great diversity of ornithologists, of their associations with their fellow workers, of their sources of funding, and of the locations of their research and publications. Yet in spite of this enormous diversity it is legitimate to speak of an international *discipline* of ornithology by 1830. For by then the study of birds had gone from an investigation of a small number of specimens by a few individuals, who devoted upon them only a portion of their scientific interest, to a scientific discipline with an enormous empirical base and a considerable number of individuals who studied it. In the process the study of birds became more specialized and more rigorous. The audience for these scientists evolved correspondingly. In the 1760's and 1770's one wrote for a general natural history reader. By the early decades of the nineteenth century an ornithologist was writing primarily for the benefit of other naturalists, and after 1830 he wrote for other specialized ornithologists. All along, of course, there were recognized experts, the number of which continually increased. These experts, however, did not constitute a profession. It was the case that the general status of the scientist did improve in the nineteenth century and that scientists came to be regarded as professionals. The organization and institutional structures of science also improved. However, the bulk of the dramatic gain in all these areas took place well after the emergence of ornithology as a scientific discipline.

Most of the key words that figure in the discussions of the professionalization of science have been conspicuously lacking in this monograph. This has not been due to a desire to eschew jargon, but reflects the fact that features such as specialized education, competency tests, codes of ethics, preemption, legitimation, etc., were not aspects of the emergence of ornithology as a discipline.[18] Although some of the factors that were important for the professionalization of science, e.g., increased potential for patronage, were also important in the history of the fragmentation of natural history into separate disciplines, it is possible to study the creation of a discipline independently from its professionalization, at least in the case of ornithology. (Indeed, it also can be pointed out that professionalization has never been complete in ornithology.[19]) There are other fields of science, such as chemistry, where the story may be quite different. For those situations, however, where the emergence of the discipline can be handled satisfactorily without reference to professionalization, it may be a great aid to isolate the two discussions. The failure to separate different historical trends partly accounts for some of the dilimmas in the current literature on nineteenth-century science. For example, Susan Faye Cannon has devoted considerable attention to the problem of characterizing the professionalization of science and of establishing criteria by which to distinguish professionals.[20] Cannon enjoys tweaking the noses of sociologists and fellow historians, who have offered definitions of a professional scientist, by showing that their definitions lead to such seemingly absurd conclusions as denying professional status to Darwin or Lyell.

Roy Porter has tried to refocus the problem by substituting the term "career" naturalist in place of "professional". According to Porter, career naturalists were those who considered science a vocation and who "became a self-sustaining, self-validating knowledge elite, guardians of expertise in their fields of intellectual endeavor".[21]

By urging that historians should, where possible, discuss the rise of disciplines separately from the professionalization of science, I do not wish to imply that the professionalization of science is a topic that historians have invented or are reading into the past. Many scientists in the nineteenth century were concerned with their professional status. Swainson, for example, was quite outspoken about it.[22] However, the preoccupation with a nomenclature to describe professional activity of the 1840's has, perhaps, obscured the more fundamental transformation of the field of natural history itself. It is not that Bonaparte was more professional than Brisson or Buffon. The reason Bonaparte's career and publications differ from Brisson or Buffon's is that Bonaparte was working within a scientific discipline that had not existed a century earlier. The new discipline was populated by a heterogeneous lot, especially if one considers the international community of ornithologists, not just the national circles. However, in spite of the diversity of ornithologists, the discipline had definable coherence. It was united by shared goals, methods, and problems, a large empirical base, avenues of communication, and recognized experts. An important implication of this monograph, then, is to strengthen the value of distinguishing between discipline and profession. In present day science it is difficult to separate the two;[23] the emergence of a new discipline is often coincident with the emergence of a new profession. But for historical analysis, especially when dealing with the period before the late nineteenth century, the distinction should be considered seriously.

An issue in the history of science that is related to the professionalization of science, and for which this monograph is relevant, is specialization in science. Everett Mendelsohn has written that:

Specialization ... seemed to take the whole man working full time. This required the new positions, the support, the organization for the advancement of science and its practitioners, the professional standards which came to mark science during the nineteenth century. It is difficult to overestimate the impact that specialization had upon the organizational and institutional structure of science. ... But specialization was only one cause, albeit an important one, among many for changes we have lumped together and called professionalization.[24]

Specialization was an important factor in the professionalization of science; however, as Mendelsohn notes, we need to be careful not to infer that by itself specialization led to professionalization. We should also not regard the significance of specialization solely in its relationship to professionalization. We have just emphasized the necessity of distinguishing the history of the profession from the history of the discipline, and we earlier noted the tie between specialization and the emergence of ornithology as a scientific discipline. What characterized the new discipline of ornithology was its concentration on one narrow domain: the birds. Initially many authors felt obliged to justify their limited field of endeavor. In one of the earliest monographs on birds Levaillant included the following explanation of the value of specialized studies:

The naturalist who wishes to encompass at once all the parts of the vast organic kingdom and to give a history of all its productions cannot − no matter how zealously he tries − go into all the details necessary for the knowledge of the animals that he treats. He can speak of them only in a superficial fashion and sometimes from the most disparate accounts. It is only the modest scholars, who limit themselves to the history of a few genera, who can hope to make known completely the species of some. Thus he who wishes to describe the regions surrounding him from the summit of a very precipitous mountain will necessarily fall into much contempt, whereas he who descends into the valley to visit a part of it will discover new objects, which would certainly have escaped the sight of the former because of the distance.

This consideration should suffice to demonstrate how much specific treatises serve the advancement of science.[25]

By 1819 such explanations were no longer necessary. William MacLeay noted in that year that the increase in information "renders it impossible for the naturalist to study in detail, more than one department of that which may be his favorite science".[26] Although some naturalists, such as Bonaparte or Yarrell, did serious research in more than one area of natural history, many took the Rev. Leonard Jenyns's advice:

for the direction of those, who may be entering on the branch of science we are here considering. We have more than once alluded to the immense field which Zoology brings before us. We have also noticed the great partiality shown by naturalists towards certain parts of that field in preference to others. Now what we recommend to such as really desire to advance its progress, — is, — that they restrict their chief attention to some given department, and, when practicable, to those particular groups which have been least studied. It is utterly out of our power to become acquainted with all the existing species of nature. The longest life, added to the enjoyment of the most favourable opportunities, will not suffice for acquiring more than a very limited knowledge of the details of their history. It must, then, be by division of labour, that we try to perfect the science, so far as human researches can perfect it.[27]

The specialization within natural history promoted detailed and rigorous research. By paying attention to limited areas, naturalists could resolve highly specific problems, correct errors, and raise new questions that previously would have been glossed over. The consequence of this research was, as we have seen, the emergence of new disciplines. Specialized societies and journals followed. The process of specialization, then, helped to define and promote the separate branches and departments of natural history, and the prodigious quantity of nineteenth-century publications in the area stand as a monument to its importance.

But there was a darker side to specialization. The most apparent drawback was the greater distance that separated the expert from the lay audience. Not that ornithology came to possess a body of esoteric knowledge that excluded all but the adepts as, say, population genetics or neurophysiology currently do. Latham is not more difficult to read or understand than Linnaeus. The increasingly narrow focus of ornithology and the growing number of specialists, however, led experts to write for other experts rather than for a general audience. There were exceptions, of course, but the process of specialization worked against the former practice of naturalists writing for the educated general reader. The recognized questions of ornithology were, after all, technical: reform of nomenclature, revision of various genera, elaboration of a natural system of classification, etc. Although a few writers continued the descriptive tradition, as established by Buffon, they were in the minority, and few had Buffon's literary skill that could engage the public's imagination.

Not that the general public was neglected! Natural history became fashionable during the nineteenth century, and there was a flood of popularizations, illustrated albums, and manuals. A few of these works were of high quality and done by outstanding naturalists. For example, William Jardine (1800–1874) collaborated with his brother-in-law, the well-known printer and engraver William Home Lizars (1788–1859), on a very ambitious project: *The Naturalist's Library*. For a decade Lizars enjoyed considerable success from this exemplary product of the "cheap literature movement" – the illustrated volumes cost only six shillings. His diligent brother-in-law provided the text for ten of the fifteen volumes on birds;[28] the other five were also done by competent ornithologists: William Swainson and P. J. Selby. The earlier *Cabinet Cyclopaedia* of Dionysus Lardner is another good example. Lardner made extensive use of Swainson's considerable talents (and need for money) for eleven volumes on

natural history. But most of the popular natural history that
saturated the Victorian market was conceived in a more sentimental or decorative fashion, in harmony with the use of glass
cases of beautiful birds for interior decoration that increased in
popularity at the same time. From the perspective of this more
vulgar natural history much of what occupied serious naturalists
appeared esoteric or excessively narrow. The satirical literature
of the time reflects such a judgement. One well-known work
that, among other topics, satirizes the mania of naturalists for
identifying new species is the witty *Histoire naturelle drolatique
et philosophique des Professeurs du Jardin des plantes, des aide-
naturalistes, préparateurs, etc., attachés à cet établissement,
accompagnée d'épisodes scientifiques et pittoresques* which relates
the following incident:

Columbus discovered America; Vasco de Gama, the route to the Indies;
Gutenberg, printing; Schwarz, gunpowder; Robert Macaire, philanthropy;
and the Academy of Moral and Impolitic Sciences, virtue. Now all of these
miserable discoverers and inventors are nothing compared to M. Blanchard,
the young friend of M. Milne-Edwards.

This young man discovered under the mantle of the *mya truncata*, not a
new animal, which would be a common thing, but rather a bizarre being, one
destined to form in the animal series a new kingdom intermediate between
the turkey and the cricket. He did not dare to draw it for fear of startling
his readers. He did not even dare to look at it which makes his discovery
even more original.[29]

The hero of the story, M. Blanchard, ultimately is disappointed
when Guérin points out that the newly discovered animal is merely
an annelid which already has been described and drawn. Although
M. Blanchard is a caricature, this satire reflects the narrowness of
much serious natural history and the general public's alienation
from it.

Aside from restricting the range of focus that naturalists considered feasible for serious research, specialization also limited the
topics deemed appropriate for discussion in natural history. In

the eighteenth century, moral, aesthetic, religious, and social comments occupied a vital role in natural history. Buffon, for example, in his ornithology extolled the social virtues of the pigeon and suggested that they would serve as good models for man to imitate.[30] George Edwards, to cite another example, included a discussion of epistemology in his work *A Natural History of Birds* and dedicated the fourth volume, "To GOD, The ONE Eternal! the Incomprehensible! the Omnipotent! Omniscient, and Almighty CREATOR of all things that exist! from Orbs immeasurably great, to the minutest Points of Matter, this ATOM is Dedicated and Devoted, with all Body and Mind, By His most resigned, Low, and humble Creature, George Edwards."[31] Pious dedications, aesthetic remarks, and didactic lessons did not totally disappear from ornithology in the first half of the nineteenth century. Natural theology inspired considerable efforts, especially in Britain and Germany, and aesthetic considerations were central to the great bird art books published. But more and more, especially after 1830, ornithology stuck closely to its clearly defined technical problems, and "digressions" were out of place or were the signs of popularization rather than the vehicle of serious communication. The internal development of ornithology and other specialties within natural history was thus away from one of its earlier major sources of motivation. Historians such as Cannon have therefore overemphasized Darwin's importance in the breakup of the alliance between science and religion in the second half of the nineteenth century.[32] It was Darwin's generation, as much if not more, that shattered the formerly close tie by restricting the scope of natural history.

Style in natural history writing suffered in a corresponding manner. It would be an oversimplification, of course, to say that natural history shifted from a literary endeavor to a technical field. The history of literary style in ornithology is complicated, and often reflects the problem undertaken rather than the talent or inclination of the author. Eighteenth-century treatises on systematics are as dull as nineteenth-century ones, whereas descriptive ornithology has had examples of literary genius even into the twentieth century. After 1830, however, the literary aspect of ornithology was incidental to the technical focus of the work. Ornithologists wrote for other ornithologists in order to communicate empirical finds or to discuss their significance. Far from attempting to depict nature's tableau or publishing, as Buffon had done, *les grandes vues*, ornithologists sought more rigorous modes of expression. Their scientific articles and monographs were for the library or study, not the boudoir or salon.

One might dismiss a concern over the greater distance between the ornithologists and their lay readers, the limitations imposed upon their topics, and the decline of the literary merit of their work as merely a sentimental lamentation about the rise of modernity. But ornithologists lost more than the ability to turn a phrase in the process of becoming practitioners of a new discipline. In reviewing the ornithological literature of the first half of the nineteenth century, what strikes the reader is the largely empirical nature of the writings. Given the pressing theoretical problems that confronted naturalists, the paucity of publications on general theoretical issues is all the more surprising. The goals of ornithology would seem to demand an extensive body of theoretical work. It is impossible, after all, to construct a natural system of classification without a coherent view of the order in nature, just as it is equally fruitless to explain geographical distribution without some high level generalizations. How can this situation be explained? Why is the balance between theoretical and empirical writings so lopsided? One might note that the increase of specimens favored empirical writings for the simple reason that there was a lot to be described, and consequently the disproportionate amount of space in journals and books devoted to descriptive or taxonomic literature merely reflects the realities of the available empirical base. The need for a catalogue of birds and a standard nomenclature were pressing affairs that led naturalists to attend to the empirical base of ornithology. The vast stores of newly discovered species, moreover, made the study timely and fruitful. One might, in addition, observe that a strongly empirical philosophy of science was prevalent at the period. Much of the historical literature about early and mid-Victorian philosophy of science stresses the empiricist tradition. Susan Faye Cannon, for example, has identified the importance of scientists such as John Herschel (1792–1871) and Alexander von Humboldt (1769–1859) in both influencing and exemplifying the practice of science in Britain.[33] Both Herschel and von Humboldt held that natural science must be built upon the rigorous examination of facts from which general laws could be inferred.[34] Recognition of a strong empiricist tradition in the philosophy of science, of course, does not solve the problem of why empiricism was prevalent at the period, and one also must keep in mind that in natural history the empiricist tradition was not universal; it was stronger in Britain than in other countries. In Germany, given the development of *Naturphilosophie*, the situation was philosophically more conducive to general theorizing, but even there the empirical literature in ornithology far outweighed the theoretical. The specialization in natural history, which took place in all the countries we've discussed, must be seen as a major contributing factor to the trend of empiricism in ornithology. The ever telescoping methodology of nineteenth-century ornithology restricted the domain of legitimate competency in scientific questions. A specialist on parrots or on humming-birds did not feel called upon or trained for commenting on the larger theoretical issues of natural history, for by definition they were outside his areas of specific or general expertise. What defined ornithology from entomology or physiology was that it focused on birds. The theoretical questions that ornithologists raised, however, went considerably beyond the scope of the study of birds, often beyond the boundaries of zoology. These issues, therefore, were outside the scope of the specialist.

For many the specialization of natural history, although valuable for the increase of rigor that it brought, appeared to be a deplorable fragmentation of science which inhibited the progress of science and the discovery of general laws of living beings. The introductory essay (1824) of the *Annales des sciences naturelles* reflects this attitude in its lament that "ornithologists, ichthyologists, entomologists, conchologists, each concentrate in the field that they cultivate and appear most often to ignore principles that their neighbors have judged must be adopted".[35] Indeed, by 1836 Neville Wood was worrying about the fragmentation of ornithology itself. He wrote: "It has long appeared to us that the greatest desideratum in Ornithology is a work devoted to the whole of the science, and including every particular which has hitherto been observed."[36] Wood was correct. After splitting off from natural history, ornithology itself was becoming specialized into subdisciplines, and the prospect of even a complete ornithology appeared remote.

Who, then, was qualified to comment on the larger issues in natural history? If naturalists increasingly restricted their domain of expertise, who could judge the validity of a claimed discovery of a general law? Although certain British philosophers expected the process of induction to generate obvious truths, the scientists were not of a single mind, nor were naturalists working with the mathematical and experimental clarity of the physical sciences on which most of the philosophy of science was patterned. The issue of justifying generalizations in natural history is not a hypothetical one. It had important implications in the most important theory advanced in natural history: the origin of species.

Darwin's theory of evolution was published two years after Bonaparte's death. The formulation of the theory of evolution by means of natural selection was one of the fundamental occurrences in the history of modern biological sciences, and not surprisingly, considerable efforts by historians of biology have been concentrated on understanding the history of the development of the theory, its reception, and its subsequent elaboration, reformulations, and extensions. Although Charles Darwin (1809–1882) had done extensive specialized research, he was not thought of as a specialist by his contemporaries. Nonetheless, his relationship to the specialization of the time is quite interesting and significant, especially his connections to ornithology. Darwin owed a lot to the new discipline of ornithology. On a technical level the empirical information and its interpretation influenced Darwin's train of thought that led to his formulation of the theory of evolution. Sandra Herbert has shown that by 1837 the "results of the professional examination of Darwin's collections were decisive"[37] in his thinking on the transmutation of species. Darwin had made some striking observations in the Galápagos Islands where he noted that in an isolated and geologically recent area there existed numerous forms of particular kinds of animals. These might have been merely varieties of distinct species, but Darwin was concerned that the various forms might turn out to be individual species. He felt that if, indeed, these various forms were good species, belief in the stability of species would be undermined. Among the most important specimens he studied in this regard were the Galápagos finches and the Galápagos mockingbirds. The "professional" who examined the specimens for Darwin was John Gould, who by then was a recognized ornithologist of the Zoological Society of London. He determined that Darwin had indeed collected specimens of distinct species. David Kohn has recently shown, in addition, that Gould's similar decision on the two rheas that Darwin collected in South America was part of the complex series of events that convinced Darwin by 1837 of the reality of transmutation.[38]

The judgements of specialists, therefore, were quite important factors in the course of Darwin's thinking.

But the new discipline of ornithology contributed to Darwin's intellectual development in a deeper fashion, for ornithology, along with the other departments of natural history, had by the late 1830's raised several issues that were germane to the origin of species. As we have seen in this monograph, the characteristics of the available empirical base, the methods of collecting and storage, and the institutional settings all tended to reinforce an interest in certain questions of systematics, distribution, and nomenclature. Although Darwin was not an experienced naturalist when he embarked on the voyage of *H. M. S. Beagle*, his experience during that expedition focused his attention on the same biological questions that his contemporaries also found of central importance. What was the difference between a variety and a species? What accounted for the distribution patterns that existed? What was the relationship between living and extinct forms? etc. Ornithology, of course, was not the sole forum for the discussions on these issues, nor was Darwin's attention to avian forms the only critical factor that led him to construct his theory of evolution. Rather, it was the process of the specialization of natural history, which we have examined by way of the history of ornithology, that led to a concentration on certain critical questions.

Darwin and the naturalists of his generation hoped to construct a picture of nature that reflected the regularities that existed, and they desired to understand those regularities. The rigor with which they approached their problems was part of a new practice in natural history. Darwin benefited from the expertise of his contemporaries, and he was able to probe his friends and acquaintances with numerous queries to which he received detailed and reliable responses. He also benefited from the experience of engaging in specialized research. His eight year travail on the barnacles has often been cited as a major step in his development. Darwin had had the good fortune to partake of the opportunities of collecting in exotic countries which had given him an exposure to field methods, to a broad array of facts in all three branches of natural history, and to the phenomena of geographical distribution. It was his work on the barnacles that engaged him in detailed taxonomic problems and in the methods and problems of systematics. Thomas Henry Huxley (1825–1895) wrote to Charles Darwin's son, Francis Darwin:

In my opinion your sagacious father never did a wiser thing than when he devoted himself to the years of patient toil which the Cirripede-book cost him.

Like the rest of us, he had no proper training in biological science, and it has always struck me as a remarkable instance of his scientific insight, that he saw the necessity of giving himself training, and of his courage, that he did not shirk the labour of obtaining it.

The great danger which besets all men of large speculative faculty, is the temptation to deal with the accepted statements of facts in natural science, as if they were not only correct, but exhaustive; as if they might be dealt with deductively, in the same way as propositions in Euclid may be dealt with. In reality, every such statement, however true it may be, is true only relatively to the means of observation and the point of view of those who have enunciated it. So far it may be depended upon. But whether it will bear every speculative conclusion that may be logically deduced from it, is quite another question.

Your father was building a vast superstructure upon the foundations furnished by the recognized facts of geological and biological science. In Physical Geography, in Geology proper, in Geographical Distribution, and in Palaeontology, he had acquired an extensive practical training during the voyage of the *Beagle*. He knew of his own knowledge the way in which the raw materials of these branches of science are acquired, and was therefore a most competent judge of the speculative strain they would bear. That which he needed, after his return to England, was a corresponding acquaintance with Anatomy and Development, and their relationship to Taxonomy – and he acquired this by his Cirripede work.

Thus in my apprehension, the value of the Cirripede monograph lies not merely in the fact that it is a very admirable piece of work, and constituted a great addition to positive knowledge, but still more in the circumstance that it was a piece of critical self-discipline, the effect of which manifested itself in everything your father wrote afterwards, and saved him from endless errors of detail.[39]

Darwin, then, had the arduous training of a specialist in zoology as well as an impressive range of knowledge in geology and distribution. During the decades that he formulated his theory to account for the origin of species, he was working with the materials and methods of a highly specialized, rigorous natural history. He was pursuing a question that his contemporaries judged central, and he was working with the rigor they deemed appropriate. The empirical data available to him was vast due to the expansion of natural history in the late eighteenth and early nineteenth centuries.

But the specialization of natural history that contributed so much to his work also hindered him. What Darwin had constructed was not a theory limited to a single discipline. The theory of evolution, in fact, cut across all three branches of natural history: zoology, botany, and geology. At a time when many of his peers were writing monographs on genera, Darwin was elaborating a synthetic system that derived its force from its ability to explain the facts of natural history and to integrate them. His theory was not inductive, as that term was understood in mid-nineteenth-century Britain, and he was later criticized there on that count. But much more serious than the fact that Darwin was employing a complex philosophy of science that didn't fit the simple models of nineteenth-century philosophers, was the problem of the lack of an appropriate forum for discussion or an authoritative body capable of commenting on the theory. Not that people didn't comment! Except for in France, lively debates ensued in countries where a scientific community existed.[40] Natural history, however, lacked the conceptual unity for a focused discussion. Instead of a single natural history there existed by the mid-nineteenth century disciplines and sub-disciplines, each with its own limited concerns and specialized methods. Even though some naturalists might do research in more than one discipline, when they published they usually did so within the confines of a particular one. The main exception was the work done on regional faunas which cut across several disciplines. Even these, however, might be considered as constituting a specialized genre within natural history and these were very limited conceptually. Comparative anatomy and physiology by this time were quite distinct from natural history, and harbored their own aspirations of discovering the laws regulating living beings. By contrast with the majority of work being done, Darwin formulated a general theory that potentially encompassed all of natural history and held implications for the rest of the life sciences. Who was able to comment with *authority* on such an intellectual construct? The very specialization that helped make the theory possible inhibited any meaningful discussion of it. Not surprisingly the history of the debate over evolution theory reveals a long and complex development. Even if we ignore the tangential issues of the philosophical, social, and religious implications, the theory generated considerable argument, confusion, and dilemmas. Historians of biology have only begun to explain this disordered state of affairs. The initial situation was compounded by numerous factors: the lack of an adequate genetics, contradictions with accepted views in other disciplines, incomplete information, etc. When the story is finally constructed this episode in the history of science will certainly be among the most interesting and illuminating. Part of that history will have to assess the impact of specialization on the reception of Darwin's theory.

The period covered in this monograph on the emergence of ornithology as a scientific discipline properly ends before the debate over Darwinian evolution commences, thus little that is conclusive can be drawn from it. Nonetheless this study does suggest questions that might be fruitful. It clearly underscores the necessity of evaluating the extent to which the debate was complicated by the specialization of natural history, and it indicates that historians should study how naturalists within different specialties reacted to evolution theory. Alfred Newton in the excellent historical introduction to his *A Dictionary of Birds* wrote: "there was possibly no branch of Zoology in which so many of the best informed and consequently the most advanced of its workers sooner accepted the principle of Evolution than Ornithology, and of course the effect upon its study was very marked".[41] If that was so, why was it? Were the questions and assumptions in ornithology more closely allied than, say, entomology to the answers offered by evolution theory? Or, by contrast with ornithology, why was French physiology totally unaffected?[42] Should Darwin's theory be regarded as a theory of natural history rather than a theory of biology? The few studies on the comparative reception of Darwin's ideas have contrasted the reactions of the scientific community in different countries or have contrasted the popular debates. Attention to the reception in different disciplines might be equally valuable. It might well help to reveal fundamental differences among disciplines or among national schools within disciplines.[43] It might also give us a clearer insight into the standards and assumptions of the different life sciences, and might thereby aid us in understanding why some of the controversy over general topics like evolution often have been so confusing and confused.

The fragmentation of natural history, which is illustrated in the case study of the emergence of ornithology as a scientific discipline, has, then, interesting implications in a number of areas in the history of biology. It bears on the issue of professionalization, on the relationship of naturalists to the lay reader, on new limitations in natural history, and on the Darwin story. This monograph also has relevance for topics that go beyond the limits of the history of science but which are of a fundamental importance for an understanding of that history. One of the key factors in the development of ornithology was the dramatic expansion of available empirical material. We have seen that several sources contributed to that expansion. With the exception of the local European faunas, the other sources were related, directly or indirectly, to European colonialization of foreign territories. The efforts of traders, explorers, colonial naturalists, etc. all stimulated the accelerating growth of ornithological collections. Colonialization and the transformation of natural history were therefore closely linked events in the sense that to appreciate the factors that transformed natural history one needs to study the context of those factors, and of central importance was the enormous wave of colonial expansion – the "second age of Empires" – that occurred in the nineteenth century. Reciprocally, part of the story of colonialization is its ⸳effect on science. It is an aspect usually overlooked in favor of the social, political, economic, religious, and technological roots or impact of Europe's expansion overseas, and its relationship to the age of imperialism that began after 1870.[44] A closer study of the relationship of colonialization to natural history would disclose an interesting facet of the complex interaction between science and society. One cannot, for example, make a simple correlation between the level of colonial activity and the growth of ornithological collections. During the first half of the nineteenth century British expansion far surpassed the French efforts, but Paris remained the center of ornithology, and the *Muséum's* collection of birds far surpassed the British Museum. Similarly, the Berlin museum was of considerable importance although Prussia was by no means an active colonial power until very late in the century.

To understand the relationship of the process of colonialization to natural history a lot more needs to be known about its history and more subtle questions need to be explored. Why did governments support the collection of natural history specimens? Obviously there were certain economic incentives that had long been appreciated. The early Spanish explorations of the New

World brought back maize which "changed the world as much as all the bullion they carried".[45] Potatoes, peanuts, manioc, and cassava are some of the better known staples that were transferred from the New World. The growing industrialization of Britain expanded the interest in exotic substances of commercial value. Humphry Davy (1778–1829), for example, rescued the tanning industry by following a suggestion of Joseph Banks (1743–1820) that led to the discovery of a new source of inexpensive tannin in the extract of a species of mimosa which grows in Bengal and Bombay and which the East India Company could supply in sufficient quantity.[46] But a purely economic explanation is not sufficient to explain the inventories of material brought back by explorations where we find listed sizable quantities of specimens that had no foreseeable economic value at the time; thousands of exotic bird skins to cite an appropriate example. A careful study of the history of natural history collecting, from a technical and social point of view, would be of considerable value in understanding the relationship of colonialization and natural history. Governments, of course, differed in their levels and means of support for collectors. A comparative study of the British, French, Dutch, and German policies would show differences in national styles of funding and support for natural history. The British were not well organized. Private entrepreneurs, such as Leadbeater and Hugh Cuming (1791–1865), managed to assemble tremendous collections,[47] whereas the government merely gave local assistance where possible, permitted naval officers to function as expedition naturalists, and occasionally gave direct subsidies to individuals. The major responsibility for encouraging, organizing, and utilizing collections rested either in private hands or in corporate or learned societies such as the Zoological Society, the East India Company, or the Linnean Society. The French government, by contrast, in addition to encouraging naval officers to take on the added responsibility of knowledgeable collecting on major expeditions, provided the *Muséum* with an annual budget to train and equip field collectors. The professors of the *Muséum* sent their *voyageurs-naturalistes* to French colonies and the areas outside of French domination in an effort to collect on a global scale. Paris was also the seat of the Maison Verreaux, the largest private natural history supply house of the day. The Dutch, in addition to continuing their encouragement of traders and private collecting, established the well funded *Natuurkundige Commissie* to exploit the riches of its own colonial holdings which they attempted to keep as isolated as possible from contact with other European colonial powers. The German states, especially Prussia, although lacking an empire, financed important individual expeditions to areas that were open to foreign collectors.

It would be an error to assume that the quantity of material collected and the level of government support led to preeminence in natural history. Several of the potentially most valuable collections sat packed in boxes for decades, and large institutions did not always foster the most advanced work. How, then, did naturalists benefit from colonialization? It was not that colonialization led automatically to an influx of natural history specimens, but rather that naturalists were able to exploit the opportunities that colonialization made available. Areas of the globe that had been inaccessible came to be open for exploration. For numerous reasons that were discussed in chapters three and four new sources of support came available. Government expeditions sent out to survey, explore, or open up avenues of trade could easily and conveniently permit the collection of natural history specimens, and they chose to do so. The specimens brought back were in sufficient quantity so as to lead to a reorganization of the practice of natural history. The intrinsic scientific value of the specimens influenced the direction of the questions addressed within natural history and raised important theoretical problems. Colonialization, therefore, was an important and complex contributing influence in the restructuring of natural history. But to understand in detail this complicated impact, a much more careful investigation will have to be made. As was just shown, simple correlations don't hold. Only a careful study of the comparative history of collecting in the nineteenth century will demonstrate how naturalists benefited from the new opportunities opened up by colonialization.

If colonialization helped to expand the available empirical base of natural history, the growth of museums in the nineteenth century provided the environment in which these specimens were examined and discussed. Museums, public and private, were the repositories for the notes, drawings, skins, and skeletons that collectors brought back to Europe. They formed a new international network in which specimens were exchanged, extensive holdings were made available for study, and catalogues, which served as guides to the collections and as milestones of the progress of science, were compiled. No one museum was complete in the first half of the nineteenth century, and naturalists usually felt obliged to visit several for an extensive project. By the time of Gould and Bonaparte it was no longer possible to work in the style of Latham who attempted to construct complete catalogues without even crossing the English Channel.

From where did these great museums, which displaced *cabinets d'histoire naturelle* for the study of ornithology, come? In chapter four some of the diverse conditions that made museums possible were described. Although there was no common pattern, a comparative study might show that many museums owed their origins to some shared cultural conditions. For example, during the first half of the nineteenth century European states altered and reformed their educational institutions.[48] These early reforms were not, for the most part, undertaken with any intention of improving the study of natural history, but naturalists benefited nonetheless from the new opportunities that came available. In a similar way the reform movements of the early nineteenth century helped to create an atmosphere that was conducive to the foundation and support of institutions for public instruction. Supporters of museums often stressed the didactic dimension of natural history and consequently gained support even in difficult times: the Paris *Muséum* was among the very few institutions that survived the revolution intact,[49] the Museums Act in England was passed during the "hungry forties", and the Berlin museum was founded in the wake of Napoleon's occupation. The broad support for public instruction was a factor in these examples and in other cases. Although the great national museum collections didn't fully develop until late in the century, their early start was of vital significance for the development of natural history and laid the foundations for the later expansion of natural history museums.

This monograph has demonstrated the importance of museum collections for the rise of ornithology. It has, however, only superficially discussed the rise of museums. What would improve our understanding of the transformation of natural history and provide a better appreciation of how that change was part of a general cultural evolution would be a social history of natural history museums. At present it is an almost unexplored area of research. It is, however, one of the important chapters in the story of the nineteenth-century context of science.

A related issue, which equally needs to be explored in depth, is the general increase in the popularity of natural history. The growth in the quantity of active researchers, the increased size of their audience, the profusion of publications, the differentiation of the levels of natural history discourse, all attest to a flourishing of natural history in the nineteenth century.[50] There were numerous technical factors that contributed to the fashion. For example, the revolution in the printing industry brought down the cost of producing natural history works, with the result that a flood of inexpensive manuals, periodicals (serious and popular), and encyclopedias appeared. The development of lithography allowed natural history illustrations to be less expensive and more accurate. The improvement of transportation made excursions into the country more accessible; the railroad later brought down the cost. The nineteenth century wave of fashion for natural history, however, cannot be adequately explained solely by reference to technical factors, important though they may be. A social history of natural history is necessary to supply other important reasons. David Allen has pioneered this much neglected subject in his *The Naturalist in Britain* where he describes some of the important strands of the Victorian setting. He notes, for instance, the role of the pervasive Evangelicalism:

The rising tide of industrialism found itself accompanied by a new ethical code perfectly suited to its furtherance and by a concentration of energies which greatly speeded its progress. The moral and the useful became, increasingly, intertwined: pursuits like geology could be justified, in the self-same flickering of conscience, as a means of revering the earthly grandeurs of Creation — the Natural Theology as re-enunciated by Paley and now taken up in a ceaseless chant in every preface — as a means of prospering materially. Any pursuit, on the otherhand, without an ostensible core of usefulness but which all the same exercised a compulsive fascination, whether it be climbing mountains or forming collections of flowers, had to be defended from the barbs of soulless utilitarians; and the simplest way of doing this, a way which was ordinarily unanswerable, was to discover some moral content and so proclaim its edifying character — as it were, as a kind of afterthought . . . .

For the most part all that resulted from this was mere emotional re-labelling. The new middle class laid claim to the playthings of its predecessors and received them with an awed respect, hardly altering their outward appearance even though they used them inevitably in a somewhat novel fashion. Just as they acquired the taste for Gothic and redirected it from old and crumbling ruins to modern edifices, so they retained the Rousseauist view of nature and translated it into an earnest religiosity.[51]

Much of the fashion for natural history was often vulgar. Allen writes:

The tasteful exercise of sentiment, so long and generally accepted as the hallmark of a cultivated gentility, tended to lie beyond the reach of minds dulled by industrial routines or by the no less stunting effects of a too literal fundamentalism. Increasingly, it turned instead into a feigned emotion: into Sentimentality, the mere sop to fashion of those who could not or would not commit themselves in the fuller way required — a debased substitute which by reason of its very shallowness was able to travel much faster and much farther.

By the 1820's this had made its début in France . . . . Although we may deplore its effects, we must be grateful for this final surge of would-be Romanticism, for it came just at the moment when the new middle classes were succeeding to dominance and thereby helped to keep their gaze firmly on nature.[52]

Allen is quite correct in emphasizing the importance of the nineteenth-century taste for natural history, which helped to create the establishment of the popular natural history literature that abounded during the period, for it also led many individuals to pursue more seriously their interests in natural history and to make, thereby, contributions to it. There is, of course, a certain irony in the situation. For the naturalists, who were motivated by natural theology, by their zeal in collecting and in writing, contributed to the production of vast quantities of factual data and to the ever more comprehensive lists that led to the emergence of specialized disciplines where the discussion of ultimate causes was considered out of place.

The swelling of the ranks of natural history in the nineteenth century owed much to technical factors and to a wide spread religiosity in some countries. Economic motives brought the subject to the attention of many as well. Not only governments but private entrepreneurs saw the economic value in natural history: specimens for private display and enjoyment, popular natural history literature, illustrated books, museum collection purchases, feathers and furs for the fashion industry, and the prospect of unknown natural substances or products for agriculture, horticulture, or pharmacy. Charles Babbage (1792–1871) wrote of the potential economic value of natural history in 1835:

When we reflect on the way the very small number of species of plants, compared with the multitude that are known to exist, which have hitherto been cultivated, and rendered useful to man; and when we apply the same observation to the animal world, and even to the mineral kingdom, the field that natural science opens to our view seems to be indeed unlimited. These productions of nature, varied and innumerable as they are, may each, in some future day, become the basis of extensive manufactures, and give life, employment, and wealth, to millions of human beings.[53]

There were other motives for an interest in natural history that are more difficult to document. Arnold Thackray has argued that in order to understand the great proliferation of British science and of its institutions during the nineteenth century historians need to examine the social role science had in its cultural context.[54] In a well known case study on science in Manchester he tries to broaden the discussion on the relationship of the development of science to the Industrial Revolution. He notes:

The proliferation of institutions points to fundamental qualitative shifts in the meaning of science as a cultural activity. The transformation in the number, nature, and orientation of its devotees was fittingly underlined by the creation of the British Association for the Advancement of Science (1831) and of the associated neologism "scientist." Little attention has been given to the motor forces in this social and cognitive transformation of natural knowledge. The necessary preconditions for a given technical invention − or more perceptively, innovation − have too often been at the center in discussions of science in the Industrial Revolution. But this is to adopt a limited, historically unhelpful focus, for science has more to do with minds than with machines, as those familiar with problems of modernization in other cultures are beginning to perceive. That science may be an integral part of the British Industrial Revolution and yet have no direct bearing on processes of invention and innovation is a hypothesis that has not been discussed. Rather, its importance has yet to be grasped.[55]

Thackray argues that the key to understanding the popularity of institutions like the Manchester Literary and Philosophical Society rests not in any economic tie between science and industry, but rather in its function as a social institution that fostered the "social legitimation of marginal men". Science was a cultural means through which the new wealthy elite could define and express itself. It was an activity, moreover, that unlike other cultural modes — music, painting, literature — was particularly attractive to this new elite because of its didactic value, economic potential, and democratic structure.[56] How general the Manchester model was remains to be seen. Local natural history societies began to appear in significant numbers after 1820. As was noted in the last chapter, these societies performed more of an educational than research function. Yet they were of considerable importance in popularizing natural history and providing funds for curators, establishing museums, etc. A social history of those societies might strengthen Thackray's suggestion that science played an important social function in nineteenth-century society, and might also show other aspects of the diverse support that existed for natural history. For in addition to pious promptings, economic gain, and social advantage one should not overlook the mundane consideration that natural history affords a variety of amusements and that natural history societies provided a setting for simple social activity. In an address read in 1832 to the members of the Berwickshire Naturalists' Club, the founder of the society, Dr. George Johnston, stated:

Such, Gentlemen, is a rapid indication of the results of our first year's exertions; and in my opinion, they do not discredit, but rather justify, the expectations of those who moved the institution of this club, which I doubt not, will work still more efficiently in future years. But when I estimate the advantages of our association by the acquisitions it has made to the natural history of the country, I do it great wrong; for I hold it to be more useful as affording a point of rendezvous for the naturalists of the district, where they may cultivate a mutual acquaintance; where they may talk over their common pursuit and all its incidents; where they may mutually give and receive oral information; where each may nourish his neighbor's zeal; where we may have our "careless season," and enjoy "perfect gladsomeness;" and, assuredly the good feeling and humour which have hitherto characterised, and will continue to characterise, our every meeting, vindicate me in assigning as the distinctive character of this Club, its social character.[57]

The support for natural history, then, is a subject that is closely tied to a number of issues in social history, and will only be understood in detail when more attention (including comparative studies) is focused on the roots of the fashion of natural history in the nineteenth century.

As can be seen from this brief review of the relationship of ornithology to colonialization, museum growth, and the nineteenth-century vogue of natural history, the history of the emergence of ornithology as a scientific discipline raises historical questions, the answers to which would not only solve interesting problems within the history of science but would also help to integrate the history of science into a more general history of the nineteenth century. The emergence of ornithology as a discipline, and the larger process of the transformation of natural history were closely related to political, economic, technological, and social events. Some of the connections, if not understood in detail, are in general obvious, such as the basic role that colonialization played in providing opportunities for the expansion of the available empirical base. Other relationships are more subtle, such as the motivations of those who lent support to the growth of natural history. There are, in addition, intriguing questions that this monograph suggests which must await a fuller examination of nineteenth-century thought in its cultural, social and economic contexts before being answerable. For example, in what ways was the process of specialization in natural history related to the division of labor that was salient in so many other endeavors during the nineteenth century? In manufacturing, especially, but also in medicine, government bureaucracy, etc., the division of labor was evident. To men such as Charles Babbage, who maintained an interest in reforming science as well as improving manufacturing, the division of labor was "no less applicable to mental productions than to those in which material bodies are concerned".[58] In natural history proper *The Edinburgh*

Journal of Natural and Geographical Science in its prospectus for a new series listed a set of directors "who have undertaken the entire direction of their several departments".[59] The editor justified this novel practice by noting that:

> The present era in the history of science, has been well named by Cuvier the epoch of the *division of labour*, the usefulness of which, first taught by the Arts, is now duly appreciated in the different branches of Science. Impressed with the importance of this method of prosecuting the dissemination, as well as the acquisition, of knowledge, the Editor has sought for and obtained the co-operation of these individuals whose distinguished names appear in the list of Directors.[60]

Another question that must await more detailed research but which is equally vexing is in what specific ways were the growth and transformation of natural history related to the general acceleration in Europe of the economy, the expansion of the transportation system, and the demographic increase, i.e., the factors often cited as the background for the Industrial Revolution?[61] The relationship of the transformation of natural history to the Industrial Revolution is a particularly interesting problem, for unlike the history of the relationship of the physical sciences to the Industrial Revolution, the subject is not so easily dominated by appeals to simple economic factors, and the path is open to more complex analyses of the relationship of science and society.

The emergence of ornithology as a scientific discipline in the late eighteenth and early nineteenth centuries is a story that stands as a nexus of several different issues ranging from technical problems in ornithology to broad interpretations of cultural change. In this monograph I have presented what appears to me to be the central lines of development in the history of that event. The episode began when two mid-eighteenth century studies set a new standard for the study of birds. The subsequent growth of the available empirical base and the development of bird collections provided new data and helped determine the questions

that led to a more rigorous and more specialized study. A growing
interest in nature and an increased level of support for natural
history encouraged the expansion of ornithology, so that by the
1830's there existed a recognizable scientific discipline. In this
chapter I have tried to indicate some of the wider issues to which
that history relates. As a case study this monograph suggests that
the general interpretations of the history of natural history in the
eighteenth and nineteenth centuries are too narrowly conceived,
and that we need to expand them considerably. This history of
ornithology also bears on our understanding of the professionalization and specialization of science in the nineteenth century.
Tangentially it suggests new dimensions for the Darwin industry.
On a more abstract plane this study points to the need to integrate
the history of science with general history. For to understand
the events that led to the emergence of ornithology as a discipline
one needs to examine, more closely than has been done, the tie
of natural history with colonialization, the rise of museums, and
the place of natural history in nineteenth-century culture. Such
research would not only help to explain why and how natural
history was transformed, but would add a new dimension to
historical topics that are usually discussed with little reference
to the history of science.

NOTES

PREFACE

[1] The emergence of geology, which was included under natural history in the eighteenth century, has been treated by a few historians, most notably Roy Porter, *The Making of Geology. Earth Science in Britain 1660–1815*, Cambridge, Cambridge University Press, 1977.

[2] Gerald Lemaine, Roy MacLeod, Michael Mulkay, and Peter Weingart, eds., *Perspectives on the Emergence of Scientific Disciplines*, The Hague, Mouton, 1976.

INTRODUCTION

[1] General surveys of the period, when they discuss science at all, tend naively to focus on its "positive achievements" or uncritically state how increased support became available due to science's "obvious" tie to industry. Even E. J. Hobsbawm, whose grasp of the interconnectedness of historical events is unusually broad, rather timidly skirts the subject of science. See his survey of the period, *The Age of Revolution*, London, Abacus, 1977.

[2] The general approach is not new, of course. It is well exemplified by Stephen Mason, *Main Currents of Scientific Thought*, London, Routledge and Kegan Paul, 1953. Recent and good detailed examples are Maurice Crosland, *Gay Lussac, Scientist and Bourgeois*, Cambridge, Cambridge University Press, 1978, and Morris Berman, *Social Change and Scientific Organization. The Royal Institution, 1799–1844*, Ithaca, Cornell University Press, 1978.

[3] A notable exception is the excellent study by David Allen, *The Naturalist in Britain. A Social History*, London, Allen Lane, 1976. A few studies of the popular natural history of this period have looked at the social context in which they appeared. See for example Susan Sheets-Pyenson, "War and Peace in Natural History Publishing: *The Naturalist's Library*, 1833–1843", *Isis*, 1981, 72(261):50–72.

[4] This point is discussed in William Coleman's, *Biology in the Nineteenth Century: Problems of Form, Function, and Transformation*, New York, Wiley, 1971, and Joseph Schiller's two studies, *Physiology and Classification*,

159

Paris, Maloine, 1980, and *La notion d'organisation dans l'histoire de la biologie*, Paris, Maloine, 1978. For a discussion of the general features of the changes in natural history and its relation to the changes in physiology, see my article, "The Transformation of Natural History in the Nineteenth Century", *Journal of the History of Biology*, 1982, **15**(1):145–152.

5 Although this was the view held at the time, for example, the first edition of the *Encyclopedia Britannica*, Edinburgh, Bell and Macfarquhar, 1771, Vol. 3, p. 362, defined natural history as "that science which not only gives compleat descriptions of natural productions in general, but also teaches the method of arranging them", it can be argued that natural history was a broader enterprise in the late eighteenth century. See my "Research Traditions in Eighteenth-Century Natural History", *Atti del Convegno di Studi Lazzaro Spallanzani e la Biologia del '700 Esperimenti Teorie Instituzioni*, in press.

6 Gottfried Treviranus, *Biologie oder Philosophie der lebenden Natur für Naturforscher und Aertze*, Göttingen, Röwer, 1802, Vol. 1, p. 4. For a discussion of his concept see B. Hoppe, "Le concept de biologie chez G. R. Treviranus", in Joseph Schiller (ed.), *Colloque internationale "Lamarck"*, Paris, Blanchard, 1971, pp. 199–237.

7 Jean-Baptiste Lamarck, *Hydrogéologie ou Recherches sur l'influence qu'ont les eaux sur la surface du globe terrestre; sur les causes de l'existence du bassin des mers, de son déplacement et de son transport successif sur les différens points de la surface de ce globe; enfin sur les changemens que les corps vivans exercent sur la nature et l'état de cette surface*, Paris, chez l'auteur, 1802, p. 8. For a discussion of Lamarck's ideas see: Joseph Schiller, "Physiologie et classification dans l'oeuvre de Lamarck", *Histoire et Biologie*, 1969, 2:35–57, Richard Burkhardt, Jr., *The Spirit of System. Lamarck and Evolutionary Biology*, Cambridge, Mass., Harvard University Press, 1977, and Leslie Burlingame, "Lamarck's Theory of Transformation in the Context of His Views of Nature, 1776–1809", Ph.D., Cornell University, 1973.

8 See his classic study, Arthur O. Lovejoy, *The Great Chain of Being. A Study of the History of an Idea*, Cambridge, Mass., Harvard University Press, 1936.

9 Michel Foucault and Wolf Lepenies both discuss a general epistemological shift. See Michel Foucault, *Les mots et les choses. Une archéologie des sciences humaines*, Paris, Gallimard, 1966 and Wolf Lepenies, *Das Ende der Naturgeschichte. Wandel kultureller Selbstverständlichkeiten in den Wissenschaften des 18. un 19. Jahrhunderts*, Munich, Hansen, 1976. Jacques Roger discusses the literature on this period in his "The Living World" in G. S. Rousseau and Roy Porter (eds.), *The Ferment of Knowledge. Studies in the Historiography of Eighteenth-Century Science*, Cambridge, Cambridge University Press, 1980, pp. 255–283.

10 Nathan Reingold has recently pointed out the growing trend in the history of science away from such dichotomies as internal-external approaches and towards a more integrated conception of the history of science. See Nathan Reingold, "Through Paradigm-Land to a Normal History of Science", *Social Studies of Science*, 1980, **10**:475–496.

11 See note 3.

12 A list of important articles, bibliographies, and books would be too extensive to include. Among the more recent classics are: Jean Anker, *Bird Books and Bird Art. An Outline of the Literary History and Iconography of Descriptive Ornithology*, Copenhagen, Levin & Munksgaard, 1938; W. H. Mullens and H. K. Swann, *A Bibliography of British Ornithology from the Earliest Times to the End of 1912*, London, Macmillan, 1917; Claus Nissen, *Die illustrierten Vogelbücher*, Stuttgart, Hiersemann, 1953; René Ronsil, "L'art français dans le livre d'oiseaux, Elements d'une iconographie ornithologie français", *Mémoires du Muséum National d'Histoire Naturelle*, 1957, ser. A, **15**(1); and John Todd Zimmer, "Catalogue of the Edward E. Ayer Ornithological Collection", *Field Museum of Natural History*, Zoological Series, 1926, *Vol. 16*.

13 Maurice Boubier, *L'Evolution de l'ornithologie,* Paris, Félix Alcan, 1925.

14 Erwin Stresemann, *Die Entwicklung der Ornithologie*, Berlin, F. W. Peters, 1951. The English translation, *Ornithology from Aristotle to the Present*, Cambridge, Mass, Harvard University Press, 1975, has, in addition, a bibliographic essay written by Ernst Mayr on the history of American ornithology. It lacks, unfortunately, the original plates and the bibliography.

CHAPTER I

1 André Bourde, *Agronomie et agronomes en France au XVIIIe siècle*, Paris, S.E.V.P.E.N., 1967, Vol. 2. pp. 891–893.

2 *E.g.*, [Menon], *Traité historique et practique de la cuisine*, Paris, Bauche, 1758, Vol. 1, pp. 260–446.

3 See Sonia Roberts, *Bird-keeping and Bird Cages. A History*, Newton Abbot, David & Charles, 1972, for an interesting account of bird keeping.

4 *E.g.*, see J. M. Chalmers-Hunt, *Natural History Auctions 1700–1972. A Register of Sales in the British Isles*, London, Sotheby Parke Bernet, 1976, and Frits Lugt, *Repertoire de ventes publiques*, La Haye, Nijhoff, 1933–53.

5 Richard Altick in his book *The Shows of London*, Cambridge, Mass., Harvard University Press, 1978, describes the popularity of automata. Some of the automata were more than frivolous toys and had some serious scientific intent. See, for example, David M. Fryer and John C. Marshall, "The Motives of Jacques de Vaucanson", *Technology and Culture*, 1979, **20**(2):257–269.

[6] Andre Blum, *Les modes au XVIIe et au XVIIIe siècle*, Paris, Hachette, 1928, p. 80. Translations by the author unless otherwise noted.

[7] The word "ornithologie", which had not appeared in the third edition of the *Dictionnaire de l'Académie Françoise* (1740) debuted in the fourth edition (1762). See *Dictionnaire de l'Académie Françoise*, Paris, Brunet, 1762, Vol. 2, p. 268.

CHAPTER II

[1] The most extensive study of Brisson is Constant Merland, "Mathurin-Jacques Brisson", *Biographies vendéennes*, Nantes, Forest et Grimaud, 1883, Vol. 2, pp. 1–47. The article by René Taton in the *Dictionary of Scientific Biography* is a good summary of what is known of Brisson's life and work.

[2] For a discussion of Réaumur and his collection see Jean Torlais, *Réaumur. Un esprit encyclopédique en dehors de l'Encyclopédie*, Paris, Desclée de Brouwer, 1936.

[3] Réaumur started his bird collection in 1743. See Maurice Trembley (ed.), *Correspondance inédite entre Réaumur et Abraham Trembley*, Genève, Georg, 1943, p. 187.

[4] René-Antoine Ferchault de Réaumur, "Divers Means for preserving from Corruption dead Birds, intended to be sent to remote Countries, so that they may arrive there in good Condition. Some of the same Means may be employed for preserving Quadrupeds, Reptiles, Fishes, and Insects, by M. de Réaumur, F. R. S. and Memb. Royal Acad. Sc. Paris. translated from the French by Phil. Hen. Zollman, Esq; F. R. S.", *Philosophical Transactions of the Royal Society*, 1748, 45:305. This contemporary translation, which retains the style of time, is more widely available than the original and now quite rare pamphlet by Réaumur: *Différens moyens d'empêcher de se corrompre les oiseaux morts qu'on veut envoyer dans des pays éloignez et de les y faire arriver bien conditionnez. Quelques-uns de ces mêmes moyens peuvent être aussi employez pour conserver des quadrupèdes, des reptiles, des poissons et des insectes*, n.p., n.p., n.d. For a detailed discussion of the history of taxidermy and its relationship to the development of ornithology, see my article "The Development of Taxidermy and the History of Ornithology", *Isis*, 1977 68(244):550–566.

[5] Brisson's careful descriptions and Martinet's illustrations of Réaumur's collection have preserved the knowledge of this collection well after most of the original specimens have disappeared. Unfortunately, many contemporary collections, such as Sir Hans Sloane's, which went to form part of the British Museum, sank into decay before any careful record was made.

[6] For a good discussion and bibliography on natural history cabinets see: Yves Laissus, "Les Cabinets d'histoire naturelle", in René Taton (ed.), *Enseignement et diffusion des sciences en France au XVIIIe siècle*, Paris, Hermann, 1964, pp. 659–712.

[7] Mathurin-Jacques Brisson, *Le Regne animal divisé en IX classes*, Paris, Bauche, 1756, p. iii.

[8] Mathurin-Jacques Brisson, *Ornithologie ou Méthode contenant la division des oiseaux en Ordres, Sections, Genres, Espèces & Leurs Variétés*, Paris, Bauche, Vol. 1, p. xviii.

[9] In spite of the unparalleled resources Brisson drew upon, he was not able to observe all the species he desired to include in his ornithology, and therefore, for completeness, he reluctantly had to rely, in a number of cases, on other authors. However, he scrupulously informed his reader whether or not he had actually seen the bird in question, and if he was describing it from a complete or partial specimen. This extensive bibliography shows that where he relied on other authors his research was comprehensive.

[10] Erwin Stresemann commented on this in his *Die Entwicklung*, p. 55.

[11] Brisson gave eighty new generic names, the above refer to new groups. For an excellent discussion see J. A. Allen's "Collation of Brisson's Genera of Birds with those of Linnaeus", *Bulletin of the American Museum of Natural History*, 1910, 28:317–335.

[12] Brisson, *Ornithologie*, Vol. I, p. xv.

[13] *Ibid.*, Vol. I, p. 308.

[14] Among the works that explicitly use Brisson are John Latham, *A General Synopsis of Birds*, London, Benjamin White, 1781; P. J. E. Mauduyt de la Varenne "Ornithologie", *Encyclopédie méthodique*, Paris, Panckoucke, [1783], Vol. 1; Thomas Pennant, *The British Zoology*, London, Cymmrodorion Society, 1761–6; Pierre Sonnerat, *Voyage aux Indes orientales et à la Chine, fait par ordre du Roi, depuis 1774 jusqu'en 1781*, Paris, auteur, 1782.

[15] For literature on Buffon see E. Genet-Varcin and Jacques Roger, "Bibliographie de Buffon", in Jean Piveteau (ed.), *Oeuvres philosophiques de Buffon*, Paris, Presses Universitaries de France, 1954, pp. 513–575. For general background on Buffon's ornithology, see: *Correspondance générale recueillie et annotée par H. Nadault de Buffon*, in J. -L. de Lanessan (ed.), *Oeuvres complètes de Buffon*, Paris, Abel Pilon, 1884–5, Vol. 13–14. [The Buffon correspondence was published separately in 1860 by Henri Nadault de Buffon, however, his edition (Paris, Hachette, 1860) is scarce. Quotations will be from the Lanessan edition which is more readily available and is the edition that has been reprinted.] Pierre Flourens, *Des manuscrits de Buffon*, Paris, Garnier, 1860; Louis Roule, *Buffon et la description de la nature*, Paris, Flammarion, 1924; Roger Heim (ed.), *Buffon*, Paris, le Muséum

National d'Histoire naturelle, 1952; Otis Fellows and Stephen Milliken, *Buffon*, New York, Twayne, 1972.

[16] Daniel Mornet, "Les enseignements de bibliothèques privées (1750–1780)", *Revue d'Histoire Littéraire de la France*, 1910, **18**:460.

[17] *Lettres inédites de Réaumur*, La Rochelle, Académie des Belles Lettres, Sciences et Arts de la Rochelle, 1886, pp. 78–79.

[18] Not much is known of Buffon's sources or his use of them. Elizabeth Anderson and Stephen Milliken have done very careful research that throws some light on the subject. *E.g.*, see Elizabeth Anderson, "La collaboration de Sonnini de Manoncourt à l'*Histoire naturelle* de Buffon", *Studies on Voltaire and the Eighteenth Century*, 1974, **120**:329–358, and "Some Possible Sources of the Passages on Guiana in Buffon's *Epoques de la Nature*", *Trivium*, 1970, **5**:72–84, 1971, **6**:81–91, 1978, **8**:83–94, 1974, **9**:70–80; Stephen Milliken, "Buffon and the British", Ph.D., Columbia University, 1965.

[19] Georges-Louis Leclerc de Buffon, *Histoire naturelle des oiseaux*, Paris, Imprimerie Royale, 1770–83. Vol. 2, pp. 1–2. Hereafter referred to as *HNO*.

[20] *Ibid.*, Vol. 7, p. 328.

[21] *Ibid.*, Vol. 1, p. iv.

[22] Georges-Louis Leclerc de Buffon, *Histoire naturelle, générale et particulière*, Paris, Imprimerie Royale, 1749–89, Vol. 1, p. 26.

[23] *Correspondance générale*, Vol. 13, p. 347. Guéneau de Montbeillard collaborated on the first volumes, abbé Gabriel Leopold Bexon on the last three.

[24] See Buffon, "De la degénération des animaux", *Histoire naturelle*, Vol. 14, p. 311–374. For a discussion of the development of Buffon's ideas on this subject, see my "Buffon and the Concept of Species", *Journal of the History of Biology*, 1972, **5**(2):259–284. Jacques Roger has carefully examined Buffon's theory and its historical context in *Les sciences de la vie dans la pensée française du XVIIIᵉ siècle*, Paris, Armand Colin, 1963.

[25] Buffon, *HNO*, Vol. I, p. vi.

[26] Ronsil, *L'art français dans le livre d'oiseaux*, p. 27.

[27] "Histoire Naturelle des Oiseaux par M. de Buffon", Paris, n.p., n.p., n.d., pp. 1–2.

[28] Buffon, *HNO*, Vol. 1, p. 394.

[29] For example his remarks about supposedly isolated phenomena in his article on the "Coucou", *HNO*, Vol. 6, p. 305–351, esp. pp. 321–322.

[30] *Ibid.*, Vol. 8, p. 115.

[31] *Ibid.*

[32] *Ibid.*, p. 467.

[33] *Ibid.*, Vol. 7, pp. 108–109.

CHAPTER III

1 The fashion of natural history in the second half of the eighteenth century and beginning of the nineteenth century has often been noted. For an interesting attempt to document this trend see Don Baesel, "Natural History and the British Periodicals in the Eighteenth Century", Ph.D., Ohio State University, 1974.

2 Gilbert White, *The Natural History of Selborne*, Harmondsworth, Penguin, 1977, p. 125 [letter #7 to Daines Barrington, Oct. 8, 1770].

3 British Museum, Egerton ms. 3147, fol. 15. For an interesting discussion of Dovaston see D. E. Allen, "J. F. M. Dovaston. An Overlooked Pioneer of Field Ornithology", *Journal of the Society for the Bibliography of Natural History*, 1967, 4(6):277–283.

4 See R. J. Cleevely, "Some Background to the Life and Publications of Colonel George Montagu (1753–1815)", *Journal of the Society for the Bibliography of Natural History*, 1978, 8(4):445–480 and Bruce Cummings, "Colonel Montagu, Naturalist", *Proceedings of the Linnean Society of London*, 1914–5:43–48.

5 Col. George Montagu, *Supplement to the Ornithological Dictionary, or Synopsis of British Birds*, Exeter, Woolmer, 1813, p. vi.

6 See Jacob Kainan, "Why Bewick Succeeded: a Note in the History of Wood Engraving", *Bulletin of the United States National Museum [Contributions from the Museum of History and Technology]*, 1959, 218:185–201.

7 Cleevely, "Some Background to the Life and Publications of Colonel George Montagu", p. 448.

8 The first volume treats quadrupeds. In the foreword to the second volume, however, he tells his reader that "my favorite branch of natural history has always been ornithology": Johann Matthaeus Bechstein, *Gemeinnützige Naturgeschichte Deutschlands nach allen drey Reichen*, Leipzig, Crusius, 1789–95, band 2, pt. 1, p. v.

9 See Frédéric Mauro, *L'Expansion Européenne (1600–1870)*, Paris, Presses Universitaires de France, 1967, which has an excellent bibliography. Also see John Dunmore, *French Explorers in the Pacific*, Oxford, Oxford University Press, 1965, and J. Holland Rose, A. P. Newton, and E. A. Benians (eds.), *The Cambridge History of the British Empire*, Cambridge, Cambridge University Press, 1929–63.

10 See Jacques Berlioz, "Les premières recherches ornithologiques françaises en Afrique du Sud", *The Ostrich*, 1959, supp. 3:300–302; Vernon Forbes, "Some Scientific Matters in Early Writings on the Cape", in A. C. Brown (ed.), *A History of Scientific Endeavor in South Africa*, Capetown, Royal Society of South Africa, 1977, p. 39; and Erwin Stresemann, *Die Entwicklung*, pp. 89–103.

[11] Erwin Stresemann, "Die brasilianischen Vogelsammlungen des Grafen von Hoffmannsegg aus den Jahren 1800–1812", *Bonner Zoologische Beiträge*, 1950, 1:43–51 and 126–143.

[12] William Burchell, *Travels in the Interior of Southern Africa*, London, Longman, Hurst, Rees, Orme, and Brown, 1822–24, Vol. I, p. v.

[13] J. A. Allen, "On the Maximilian Types of South American Birds in the American Museum of Natural History", *Bulletin of the American Museum of Natural History*, 1889, 2(3):209–276, describes the more notable specimens in the Prince's collection which was purchased by the American Museum of Natural History in 1870.

[14] Published as one of the biographical sketches in *Lardner's Cabinet Cyclopedia*: William Swainson, *Taxidermy, Bibliography, and Biography*, London, Longman, Orme, Brown, Green and Longmans, and John Taylor, 1840, pp. 338–352. Additional information can be found in D. J. Galloway, "The Botanical Researches of William Swainson F. R. S., in Australia, 1841–1855", *Journal of the Society for the Bibliography of Natural History*, 1978, 8(4): 369–379.

[15] The Linnean Society has five large volumes of Swainson's correspondence which are very rich in information about his life and the state of science during the period. See "Catalogue of the Swainson Correspondence in the Possession of the Linnean Society", *Proceedings of the Linnean Society of London*, 1899–1900:25–61.

[16] Swainson, *Taxidermy*, p. 345.

[17] Archives nationales, (Paris) AJ[15] 565, folder 5. The Archives nationales has a rich collection of manuscripts concerning this project, which would make a good subject of serious inquiry. See especially Archives nationales, AJ[15] 240 and 565.

[18] The results of these expeditions were described in the volumes of the *Mémoirs* and *Annales* of the *Muséum*.

[19] See Erwin Stresemann, "Die Enwicklung der Vogelsammlung des Berliner Museums unter Illiger und Lichtenstein", *Journal für Ornithologie*, 1922, 10:498–503 and his "Der Naturforscher Friedrich Sellow († 1831) und sein Beitrag zur Kenntnis Brasiliens", *Zoologischen Jarbücher* (Abteilung für Systematik, Ökologie und Geographie der Tiere), 1948, 77(6):401–425.

[20] See Agatha Gijzen, *'s Rijksmuseum van Natuurlijke Historie 1820–1915*, Rotterdam, Brusse, 1938.

[21] *Ibid.*, p. 91.

[22] For an interesting discussion of the Portuguese expeditions and an account of the transfer to Paris see William Joel Simon, "Scientific Expeditions in the Portuguese Overseas Territories, 1783–1808; the Role of Lisbon in the Intellectual-Scientific Community of the Late Eighteenth Century", Ph.D., The City University of New York, 1974.

23 The library of the Muséum National d'Histoire naturelle in Paris has an important collection of Verreaux manuscripts which could, along with the large Verreaux dossier in the Archives nationales (Paris), form the basis of a serious study of this interesting and important family.

24 Agnes Beriot, *Grand voiliers autour du monde: Les voyages scientifiques 1760–1850*, Paris, Port Royal, 1962, p. 81. In addition to this excellent study see Beriot's "Essai sur les sources documentaires concernant les voyages de circumnavigation entrepris par la Marine Française", Diplôme, l'Institue des Techniques de la Documentation, Paris, 1958, for an extensive bibliography and list of manuscript sources.

25 See J. J. H. de Labillardière, *Relation du voyage à la recherche de LaPérouse, fait par ordre de l'Assemblee Constituante, pendant les années 1791, 1792, et pendant la 1ere et la 2e année de la République Française*, Paris, Jansen, 1800. The scientific advice given to the Entrecasteaux expedition was very detailed and gives a good picture of scientific collection of the time. See Bibliothèque de Muséum National d'Histoire naturelle, ms. 46.

26 A. L. Jussieu, "Notice sur l'expédition a la nouvelle-hollande, Entreprise pour des recherches de Géographie et d'Histoire naturelle", *Annales du Muséum National d'Histoire naturelle*, 5:7.

27 *Ibid.*, p. 10.

28 Maurice Zobel, "Les naturalistes voyageurs Français et les grands voyages maritimes du XVIII e et XIX e siècle", Doctorat en Medecine, Faculté de Medecine de Paris, Paris, 1961, p. 35.

29 Dunmore, *French Explorers*, p. 228.

30 René Lesson and Prosper Garnot, *Voyage autour du Monde, exécuté par Ordre du Roi, sur la Corvette de Sa Majesté, La Coquille, pendant les années 1822, 1823, 1824, et 1825 . . . Zoologie*, Paris, Bertrand, 1825–1830, Vol. 1, p. ii.

31 Quoted in Louis de Freycinet, *Voyage autour du monde . . . exécuté sur les corvettes de S. M. l'Uranie et La Physicienne, pendant les années 1817, 1818, 1819, et 1820*, Paris, Pillet, 1825, Vol. 1, p. xxxiv.

32 See for example the careful study by Elsa Allen, "The History of American Ornithology before Audubon", *Transactions of the American Philosophical Society*, 1951, 41(3):385–591.

33 Thomas Horsfield and Frederic Moore, *A Catalogue of the Birds in the Museum of the Hon. East-India Company*, London, W. Allen, 1854–8, Vol. 1, p. iii. For a brief discussion of this collection, which ultimately went to the British Museum (Natural History) see Charles Cowan, "Horsfield, Moore, and the Catalogues of the East India Company Museum", *Journal of the Society for the Bibliography of Natural History*, 1975, 7(3):273–284, and Mildred Archer, *Natural History Drawings in the India Office Library*, London, Commonwealth Relations Office, 1962.

[34] Martin Montgomery (ed.), *The Despatches, Minutes, and Correspondence, of the Marquess Wellesley, K. G. during his Administration in India*, London, W. Allen, 1836, Vol. 4, pp. 674–676.

[35] For example, General Hardwicke's correspondence, preserved in the British Museum (Add. ms. 9869) contains complaints of the "shabby manner in which I have been treated". (Letter of Aug. 16, 1822 to A. Macleay, fol. 102).

[36] See Barbara Beddall, " 'Un Naturalista Original': Don Félix de Azara, 1746–1821", *Journal of the History of Biology*, 1975, 8(1):15–66.

[37] See Paul Fournier, *Voyages et découvertes scientifiques des missionnaires naturalistes Français à travers le monde pendant cinq siècles XVe a XXe siècles*, Paris, Lechevalier, 1932.

[38] Linnean Society, Swainson Correspondence, letter of William Burchell to Swainson, Sept. 27, 1819.

[39] See Georges Cuvier, "Catalogue des préparations anatomiques laissées dans le cabinet d'anatomie comparée de Muséum d'Histoire Naturelle, par G. Cuvier", *Nouvelles Annales du Muséum*, 1833, 2:417–508, which lists two thousand, four hundred and fifty-two bird preparations.

[40] *Catalogue of the Contents of the Museum of the Royal College of Surgeons in London*, London, Warr, 1831, pp. 172–211.

[41] Although some of this literature, *e.g.*, on breeding pigeons, was looked down upon by many "serious" naturalists, it often was of considerable scientific value. See, James Second, "Nature's Fancy: Charles Darwin and the Breeding of Pigeons", *Isis*, 1981, 72(262):163–186.

CHAPTER IV

[1] An example of one of the most splendid of these can be glimpsed from the sale catalogue of Margaret Cavendish Harley, Duchess Dowager of Portland: *A Catalogue of the Portland Museum, lately the property of the Duchess Dowager of Portland, Deceased: Which will be sold by auction, by Mr. Skinner & Co.*, n.p., n.p., 1786.

[2] Naturalists usually had access to these collections, however. Latham, for example, refers to specimens he consulted in the Lever, Portland, Banks, and Tunstall collections.

[3] See E. Mendes da Costa, "Notes and Anecdotes of Literati, Collectors, &c. from a ms. by the late Mendes de [sic] Costa, and Collected between 1747 and 1788", *The Gentleman's Magazine*, 1812, (1):205–207 and 513–516.

[4] A. J. Desallier d'Argenville, *La Conchyliologie ou Histoire naturelle des coquilles ... Troisième édition par MM. de Favanne de Montcervelle père et fils*, Paris, DeBure, 1780, p. 193.

[5] British Museum, Add. ms. 28540, fol. 156. The nine volumes of da Costa's correspondence in the British Museum (Add. mss. 28534–28546) provide an excellent picture of natural history collecting in the second half of the eighteenth century.

[6] *Ibid.* See Altick, *The Shows of London*, for a description of the popularity of collections.

[7] W. H. Mullens, "Some Museums of Old London. I. The Leverian Museum", *The Museums Journal*, 1915, 15:123–129 and 162–172.

[8] Quoted in Mullens, "Some Museums of Old London, I.", p. 126.

[9] William Jerdan, *Men I have Known*, London, Routledge, 1866, pp. 70–71. Also see W. H. Mullens, "Some Museums of Old London. II. William Bullock's London Museum", *The Museums Journal*, 1917, 17:51–56, 132–137, and 180–187.

[10] After his death the collection was purchased by the Newcastle Literary and Philosophical Society and later formed the nucleus of the Hancock Museum. See Russell T. Goddard, *History of the Natural History Society of Northumberland, Durham, and Newcastle Upon Tyne 1829–1929*, Newcastle-upon-Tyne, Reid, 1929, pp. 12–57. George Townshed Fox in his *Synopsis of the Newcastle Museum*, Newcastle, Hodgson, 1827, p. vi, commented on the historical significance of the Tunstall collection when he noted that the catalogue was valuable "as enabling students in Zoology to compare and identify the actual specimens of many subjects which have become, in some degree classical, by their having served for the original descriptions and delineations of authors; particularly of Mr. Pennant in his various works, Brown in his 'Illustrations of Zoology', Dr. Latham in his 'Synopsis of Birds', Col. Montagu in his 'Ornithological Dictionary'". He could have added the name Bewick, as well.

[11] See Jacques Berlioz, "Les collections ornithologiques du Muséum de Paris", *L'Oiseau*, 1938, (2):237–260, and J. P. F. Deleuze, *Histoire et description du Muséum Royal d'histoire naturelle*, Paris, Royer, 1823.

[12] Bibliothèque du Muséum National d'Histoire naturelle, ms 2528, no. 42.

[13] See F. Boyer, "Le Transfert à Paris des collections du Statholder (1792)", *Annales historiques de la Revolution française*, 1971, (205):289–404. Boyer quotes André Thouin, one of the four commissioners, who wrote of this prize: "The national collection will by this merging become the most magnificent that exists in the world, and the most useful for the progress of the natural sciences", p. 393.

[14] Although, perhaps not as much as was thought. See M. Boeseman, "The Vicissitudes and Dispersal of Albertus Seba's Zoological Specimens", *Zoologische Mededelingen*, 1970, 44(13):177–206, which suggests that some specimens were hidden and not removed to France.

[15] Geoffroy gave the following list of sources:

L'ancienne collection du Muséum d'Histoire naturelle	102
La mission en Hollande de MM. Thouin et Faujas	390
Le voyage à Cayenne de M. Richard	37
Le voyage à Cayenne de MM. Leblond et Brocheton	102
Un cabinet acquis de Madame Chénié	295
Le voyage aux Antilles de Maugé	296
Les envois (de Cayenne) de M. Martin	198
Mon voyage en Egypte	39
Le voyage au Bengale de M. Macé	135
Le voyage aux Terres-Australes de MM. Perron, Lesueur, Maugé et Levillain	403
Les envois (de l'île de France) de M. Dumont	20
Le voyage en Angleterre de M. Dufresne	36
Le voyage à Java de M. Leschenault	78
Les dons de S. M. l' IMPÉRATRICE	22
Ma mission en Portugal	275
La ménagerie du Muséum d'Histoire naturelle	170
Les correspondances de MM. Baillon	176
Mes relations et correspondances particulières	637
Total	3411

from Etienne Geoffroy Saint-Hilaire, "Sur l'accroissement des collections des mammifères et des oiseaux du Muséum d'Histoire naturelle", *Annales du Muséum*, 1809, 13:88. The "old collection" dating from 1793 was listed 102 because 361 had been replaced.

[16] The budget for 1821 includes three full-time taxidermists and eight part-time. By 1832 the budget shows six full-time. Bibliothèque du Muséum National d'Histoire naturelle, ms. 2298.

[17] See "Notice sur M. Dufresne, aide-naturaliste au Muséum", *Nouvelles Annales du Muséum d'Histoire Naturelle*, 1833, 2:357–359, and Jessie Sweet, "The Collection of Louis Dufresne (1752–1832), *Annals of Science*, 1970, 26:33–71.

[18] Jean Chaia, "Sur une correspondance inédite de Réaumur avec Artur, premier Medicin du Roy a Cayanne", *Episteme*, 1968, 2:130.

[19] See Farber, "The Development of Taxidermy".

[20] Isidore Geoffroy Saint-Hilaire, *Introduction au catalogue méthodique des collections de mammifères et d'oiseaux*, Paris, Plon, 1850, pp. iv-v, proudly notes the international use of the collection. The correspondence of most of the major zoologists of the first third of the century confirms

Geoffroy's statement. For example, William Kirby (1759–1850) wrote to Alexander Macleay in 1817, "Every part of the Museum is in beautiful order, systematically arranged, so that every student may in a moment find every object that he wants, and every facility is afforded to him that he can desire. I wish the zoological department of the British Museum was in similar order". Linnean Society, Macleay Correspondence, Letter of June 25, 1817.

21 Deleuze, *Histoire*, p. 436.

22 British Museum, Add. ms. 28544 fol. 148. Letter of March 4, 1784.

23 See Mullens, "Some Museums of Old London, I", for an account of the fate of the Lever museum.

24 See *Catalogue of the Leverian Museum*, London, Hayden, 1806.

25 See A. von Pelezin, "Birds in the Imperial Collection of Vienna obtained from the Leverian Museum", *Ibis*, 1873, 3:14–54 and 105–124.

26 See *Catalogue (Without which no Person can be admitted either to the View or Sale) of the Roman Gallery, of Antiquities and Works of Art, and of the London Museum of Natural History: (Unquestionably the Most Extensive and Valuable in Europe) at the Egyptian Hall in Piccadilly; Which will be Sold by Auction, positively without the least reserve, by Mr. Bullock*, n.p., n.p., n.d. [1819].

27 *Catalogue*, p. 4.

28 See my article, "The Type-Concept in Zoology during the First Half of the Nineteenth Century", *Journal of the History of Biology*, 1976, 9(1): 93–119.

29 John Latham, *A General History of Birds*, Winchester, Jacob and Johnson, 1821–4, Vol. 1, p. ix.

30 See *The History of the Collections in the Natural History Departments of the British Museum*, London, British Museum, 1904–6, and Albert Gunther, *A Century of Zoology at the British Museum through the Lives of Two Keepers 1815–1914*, London, Dawsons, 1975.

31 Great Britain, *Parliamentary Papers* (Commons), "Report from the Select Committee on the Condition, Management and Affairs of the British Museum", 1835–36, 1:217. (Hereafter referred to as *Report from the Select Committee*.) This two-volume report is extremely valuable in that it contains testimony from the leading figures of the period.

32 *Ibid.*, pp. 242–243.

33 See Charles Cowan, "Horsfield, Moore and the Catalogues of the East India Company Museum", *Journal of the Society for the Bibliography of Natural History*, 1975, 7(3):273–284. Although the collection originally was very impressive, many of the specimens were poorly preserved and were ruined by the time they were moved to the British Museum (1863). One better appreciates the importance of Dufresne and other successful

taxidermists when one realizes the extent of damage done to bird skins by faulty preserving.

[34] Bibliothèque du Muséum National d'Histoire naturelle, ms 2613, no. 3545, letter of July 7, 1828. For the popularity of the Zoological Society also see John Bastin, "The First Prospectus of the Zoological Society of London: New Light on the Society's Origins", *Journal of the Society for the Bibliography of Natural History*, 1975, 5(5):369–388, and "A Further Note on the Origins of the Zoological Soceity of London", *Ibid.*, 1973, 6(4):236–241. Also: Henry Scherren, *The Zoological Soceity of London*, London, Cassell, 1905.

[35] This point was made emphatically in the *Report from the Select Committee*, Vol. 1, p. 203.

[36] See H. Engel, "Alphabetical List of Dutch Zoological Cabinets and Menageries", *Bijdragen tot de Dierkunde*, 1939, 27:247–346.

[37] François Levaillant, *Histoire naturelle des oiseaux d'Afrique*, Paris, Fuchs, 1798, Vol. 1, p. 89.

[38] For its history see Agatha Gijzen, *'s Rijksmuseum*.

[39] *Report from the Select Committee*, Vol. 2, p. 184.

[40] The following discussion is drawn from Gijzen, *'s Rijksmuseum*.

[41] He refers to this in his correspondence, *e.g.*, Bibliothèque du Muséum National d'Histoire naturelle, ms. 1989, No. 904, letter of October 16, 1820 to Cretzschmaer, director of the natural history museum at Frankfurt am Main. Temminck was also skilled at bird preservation, having been instructed by Levaillant, who in turn learned the art from Bécoeur.

[42] See T. G. Ahrens, "The Ornithological Collections of the Berlin Museum", *Auk*, 1925, 42:241–245, August Braun, "Das zoologische Museum", in Max Lenz (ed), *Geschichte der königlichen Friedrick-Wilhelms-Universität zu Berlin*, Halle, Verlag der Buchlandlung des Waisenhauses, 1910, Vol. 3, pp. 372–389, and Gottfried Mauersberger, "Über wertvolle Stücke der Vogelsammlung des Berliner Naturkundemuseums", *Wissenschaftliche Zeitschrift der Humboldt-Universität zu Berlin*, 1970, 17:152–155. Although interesting studies have been done on the development of scientific research (mostly in the physical sciences or in physiology) in the German university, the history of the development of German natural history museums remains relatively untouched.

[43] Erwin Stresemann, "Die Entwicklung der Vogelsammlung des Berliner Museums unter Illiger und Lichtenstein", *Journal für Ornithologie*, 1922, 70:500.

[44] By 1854 Lichtenstein could report that the collection had 13,760 specimens representing 4,070 species. See M. H. K. Lichtenstein, *Nomenclator*

Avium Musei Zoologici Berlinensis, Berlin, Königlichen Akademie der Wissenschaften, 1854, p. viii.

45 [G. Heldmann], *Johann Jakob Kaup: Leben und Wirken des ersten Inspektors am Naturalien-Cabinet des Grossherzoglichen Museums 1803–1873*, Darmstadt, Selbstverlag des Verfassers, 1955, C. E. Hellmayr, "The Ornithological Collection of the Zoological Museum in Munich", *Auk*, 1928, **45**, Otto Taschenberg, "Geschichte der Zoologie und der zoologischen Sammlungen an der Universität Halle 1694–1894", *Abhandlungen der Naturforschenden Gesellschaft zu Halle*, 1894, **20**:1–177.

46 Einai Lönnberg, "The Natural History Museum (Naturhistoriska Riksmuseum) Stockholm", *Natural History Magazine*, 1933, **4**(27):77–93, Ragnar Spärck, *Zoologisk Museum i Kφbenhavn gennen tre aarhundredar*, Copenhagen, Munksguard, 1945, L. J. Fitzinger, "Geschichte des kais. kön. Hof-Naturalien-Cabinet zu Wien", *Sitzungsberichte der Kaiserlichen Akademie der Wissenschaften. Mathematisch-Naturwissenschaftliche Classe*, 1856, **21**: 433–479.

47 C. J. Temminck, *Catalogue systématique du cabinet d'ornithologie et de la collection de Quadrumanes*, Amsterdam, Sepp, 1807, gives a good idea of the extent of the collection. Also see J. A. Susanna, "Levensschets van C. J. Temminck", *Handelingen van de jaarlijksche algemeene vergadering van de Maatschappij der Nederlandsche Letterkunde te Leiden*, 1858:47–78.

48 British Museum, Egerton ms. 3147, fol. 71. Letter of Sept. 7, 1828.

49 *A Catalogue of the Menagery and Aviary at Knowsley, formed by the late Earl of Derby, K. G.*, Liverpool, Walmsley, 1851.

50 Louis Fraser, *Catalogue of the Knowsley Collections, Belonging to the Right Honourable Edward (Thirteenth) Earl of Derby, K. G.*, Knowsley, by author, 1850, p. iii.

51 *Ibid.*, pp. iii–iv. The collection ultimately went to the town of Liverpool where it served as the foundation for the very popular Derby Museum.

52 See [Jules Verreaux], *Catalogue des oiseaux de la collection de feu Mr. Le B^{on} de La Fresnaye de Falaise*, n.p., n.p., n.d., and Outram Bangs, "Types of Birds Now in the Museum of Comparative Zoology", *Bulletin of the Museum of Comparative Zoology*, 1930, **70**(4):147–426. La Fresnaye's collection was sold at public auction four years after his death. The entire collection was purchased by the city of Boston and now resides in the Museum of Comparative Zoology. See Thomas Renard, "Lafresnaye", *Auk*, 1945, **62**(2): 227–233.

53 *Catalogue des oiseaux de la collection de M. le Baron Langier de Chartrouse*, Arles, [Garcin], 1836.

54 See *Catalogue de la magnifique collection d'oiseaux de M. Le Prince*

d'Esling, duc de Rivoli, Paris, Schneider & Langrand, 1846. The entire collection of over 10,000 specimens was purchased by the Academy of Natural Sciences of Philadelphia.

[55] The famous collection of Duchesne de Lamotte still resides in Abbeville in the musée Boucher de Porthes.

[56] See J. A. Allen, "On the Maximillian Types of South American Birds". The American Museum of Natural History purchased Maximillian's collection of 4,000 mounted birds in 1870. It was an especially valuable collection due to the number of types, as described in the Allen article.

[57] Ernest Hartert, "Eine bedeutende Vogelsammlung des 18. Jahrhunderts", *Ornithologische Monatsberichte*, 1923, **31**(4):73–75 has a description of the duke's collection, Wilhelm Petry, "Eine bedeutende Vogelsammlung des 18. Jahrhunderts", *Ornithologische Monatsberichte*, 1938, **45**(5):157–162 describes the destruction of the duke's collection.

[58] See, for example, William Blanpied, "Notes for a Study on the Early Scientific Work of the Asiatic Society of Bengal", *Japanese Studies in the History of Science*, 1973, **12**:121–144, F. Brandt, "Das zoologische und vergleichend-anatomische Museum", *Bulletin de l'Académie Impériale des sciences*, 1864, **7**, supp. 2:11–28, Frank Burns, "Charles W. and Titian R. Peale and the Ornithological Section of the Old Philadelphia Museum", *Wilson Bulletin*, 1932, **44**:23–35, W. Stone, "Some Philadelphia Ornithological Collections and Collectors, 1734–1850", *Auk*, 1899, **16**:166–177, and R. F. H. Summers, *A History of the South African Museum, 1825–1975*, Capetown, Balkems, 1975.

[59] Temminck looked down upon Lichtenstein's practice and complained of it in a letter to Cretzschmar. Bibliothèque de Muséum National d'Histoire naturelle, ms. 1989, No. 931.

[60] [P. B. Duncan], *A Catalogue of the Ashmolean Museum, descriptive of the Zoological specimens, antiquities, coins, and miscellaneous curiosities*, Oxford, Oxford University Press, 1836, p. 6.

CHAPTER V

[1] See Genet-Varcin and Roger, "Bibliography", for editions.

[2] James Hall Pitman, *Goldsmith's Animated Nature. A Study of Goldsmith* (*Yale Studies in English*, Vol. **66**), New Haven, Yale University Press, 1924, p. 35. Pitman's book was reprinted in 1972 by The Shoestring Press.

[3] P. J. E. Mauduyt de la Varenne, "Ornithologie", Vol. 1, p. 405. The date of publication is complicated. See C. Davies Sherborn and B. B. Woodward, "On the Dates of Publication of the Natural History Portions of the 'Encyclopédie Méthodique'", *The Annals and Magazine of Natural History*, 1906, **17**:577–582.

4 John Latham, *A General Synopsis*, pp. i–ii.

5 *Ibid.*, p. iv.

6 See the careful study by P. J. P. Whitehead, "Zoological Specimens from Captain Cook's Voyages", *Journal of the Society for the Bibliography of Natural History*, 1969, 5(3):161–201, esp. p. 181. Also see David Medway, "Some Ornithological Results of Cook's Third Voyage", *Journal of the Society for the Bibliography of Natural History*, 1979, 9(3):315–351.

7 G. M. Mathews, "John Latham (1740–1837): an Early English Ornithologist", *Ibis*, 1931, 1(3):466.

8 For the White notebooks see Hubert Massey Whittell, *The Literature of Australian Birds*, Perth, Paterson Brokenstra PTY. Ltd., 1954, p. 29; the Cook material, Whitehead, "Zoological Specimens", and the Hardwicke material, British Museum, Add. ms. 9869, fol. 109, letter of Nov. 6, 1822, in which Hardwicke promises Latham first look at his birds, and Add. ms. 29,533, fol. 214–5, letter of May 10, 1825, in which Latham acknowledges receipt of bird drawings from Hardwicke.

9 Latham, *A General Synopsis*, p. i.

10 For a detailed and interesting comparison of Brisson and Linnaeus, see J. A. Allen, "Collation of Brisson's Genera".

11 The literature on Linnaeus is enormous. For a good discussion of his popularity and impact see Frans Stafleu, *Linnaeus and the Linnaeans. The Spreading of Their Ideas in Systematic Botany 1735–1789*, Utrecht, Oosthoek, 1971. For a clear exposition of Linnaeus's classification see James Larson, *Reason and Experience. The Representation of Natural Order in the Work of Carl von Linné*, Berkeley, University of California Press, 1971.

12 Linnean Society of London, Smith Correspondence, Vol. 23, p. 158, letter of March 8, 1821.

13 Mathews, "John Latham", p. 474.

14 Buffon, *HNO*, Vol. 1, pp. i–ii.

15 James Fisher, *The Shell Bird Book*, London, Ebury Press & Michael Joseph, 1966, p. 71.

CHAPTER VI

1 The literature on this subject is extensive. Quite useful are: Henri Daudin, *De Linné à Jussieu. Les méthodes de la classification et l'idée de série en Botanique et en Zoologie (1740–1790)*, Paris, Alcan, 1926; Emile Guyénot, *Les Sciences de la vie aux XVIIᵉ et XVIIIᵉ siècles*, Paris, Albin Michel; and P. R. Sloan, "John Locke, John Ray and the Problem of the Natural System", *Journal of the History of Biology*, 1972, 5:1–53.

[2] Louis-Jean-Marie Daubenton, "Introduction a l'histoire naturelle", *Encyclopédie méthodique. Histoire naturelle des animaux*, Paris, Panckoucke, 1782, Vol. 1, p. iii.

[3] See William Coleman, *Georges Cuvier. Zoologist*, Cambridge, Mass., Harvard University Press, 1964; and Henry Daudin, *Cuvier et Lamarck. Les classes zoologiques et l'idée de série animale (1790–1830)*, Paris, Alcan, 1926.

[4] Leonard Jenyns, "Report on the Recent Progress and Present State of Zoology", *Report of the Third Meeting of the British Association for the Advancement of Science*, 1834:143–148.

[5] *Lettres de Georges Cuvier à C. H. Pfaff, 1788–1792, sur l'histoire naturelle, la politique, et la littérature. Traduites de l'allemand par Louis Marchant*, Paris, Masson, 1858, p. 178.

[6] Georges Cuvier, *Le Règne animal distribué d'apres son organisation, pour servir de base a l'histoire naturelle des animaux et d'introduction a l'anatomie comparée*, Paris Deterville, 1817, Vol. 1, p. xxii.

[7] Coleman, *Georges Cuvier*, p. 67.

[8] Henri-Marie Decrotay de Blainville, "Sur l'emploi de la Sternum et de ses annexes pour l'établissement ou la confirmation des familles naturelles parmi les oiseaux", *Journal de Physique, de chemie, d'histoire naturelle et des arts*, 1821, 92:185–186. Blainville's memoir had been read at the *Académie des sciences* in 1815, but was not published until 1821 in the *Journal de physique*. It was known and discussed, although Blainville never completed the details. His work inspired Dr. F. J. L'Herminer, who was given access to the comparative anatomy collection by Cuvier, and who published a detailed memoir on the subject, "Recherches sur l'appareil sternal des oiseaux, considéré sous le double rapport de l'ostéologie et de la myologie; suivies d'un Essai sur la distribution de cette classe de vertébrés, basée sur la considération du sternum et de ses annexes", *Mémoires de la Société Linnéenne de Paris*, 1827:1–93. A second edition, with an introduction was published separately under the same title the following year (Paris, Desbeausseaux, 1828). The main features of L'Herminer's sytem which was the same as Blainville's, were that he established separate orders for the parrots, the ostriches and cassowarys, and the pigeons, bringing the number of orders from 6 to 9.

[9] Louis-Pierre Vieillot, "Ornithologie", *Nouveau dictionnaire d'histoire naturelle*, Paris, Deterville, 1818, Vol. 24, p. 69.

[10] For an interesting discussion of Vieillot's ornithology see Paul Oehser, "Louis Jean Pierre Vieillot (1748–1831)", *Auk*, 1948, 65(4):568–576 and Georges Olivier, *Un grand ornithologiste normand Louis-Pierre Vieillot. Sa Vie–Son Oeuvre*, Fécamp, Durand, 1965, the printed version of a lecture given by Olivier at the Académie des sciences, belles-lettres et arts de Rouen in 1961 which draws heavily on Oehser.

11 Caroli Illigeri, *Prodromus Systematis Mammalium et Avium*, Berlin, Salfeld, 1811.

12 Johann Karl Wilhelm Illiger, *Versuch einer Systematischen vollständigen Terminologie für das Thierreich und Pflanzenreich*, Helmstädt, Fleckeisen, 1800. Ernst Mayr called attention to Illiger's work in "Illiger and the Biological Species Concept", *Journal of the History of Biology*, 1968, 1(2): 163–178. More recently Phillip Sloan has discussed Illiger's importance in "Buffon, German Biology, and the Historical Interpretation of Biological Species", *British Journal for the History of Science*, 1979, 12(41):109–153. Timothy Lenoir gives an excellent description of the Göttingen School where Blumenbach worked in "The Göttingen School and the Development of Transcendental Naturphilosophie in the Romantic Era", *Studies in History of Biology*, 1981, 5:111–205.

13 In his monograph on pigeons Temminck divides the "tribe" of pigeons into three families and states "This subdivision is founded principally on the habits and the type of food proper to the birds which comprise these families". C. J. Temminck, *Histoire naturelle générale des pigeons et des Gallinacés*, Amsterdam, Sepp, 1813, Vol. 1, p. 32.

14 C. J. Temminck, *Observations sur la classification méthodique des oiseaux, et remarques sur l'analyse d'une nouvelle ornithologie élémentaire par L. P. Vieillot*, Amsterdam, Dufour, 1817, p. 5.

15 In the introduction to the second edition of his *Manuel*, Temminck wrote that he had consulted all the major European collections with the exceptions of Madrid and Saint-Petersburg. C. J. Temminck, *Manuel d'Ornithologie, ou Tableau systématique des oiseaux qui se trouvent en Europe; précéde d'une analyse du système général d'ornithologie, et suivi d'une table alphabétique des espèces*, Paris, Dufour, 1820, Vol. 1, p. ix. He also did some field work, although more for aquatic animals than birds.

16 *Ibid.*, pp. i–ii.

17 Alfred Newton, *A Dictionary of Birds*, London, Adam and Black, 1893–1896, p. 21.

18 See Ronsil, "L'art français", Nissen, *Die illustrierten Vogelbücher*, and Anker, *Bird Books and Bird Art*.

19 Ronsil, "L'art français", p. 33.

20 J. B. Audebert and L. P. Vieillot, *Histoire naturelle et générale des Colibris, oiseaux-mouches, jacamars et promerops*, Paris, Desray, 1802, p. 3.

21 Ronsil, "L'art français", p. 37.

22 Louis-Pierre Vieillot's, *Histoire naturelle des plus beaux Oiseaux chanteurs de la zone torride*, Paris, Dufour, 1805 and *Histoire naturelle des Oiseaux de l'Amérique septentrionale contenant un grand nombre d'espèces décrites ou figurées pour la première fois*, Paris, Desray, 1807, are important iconographically.

23 Temminck, *Histoire naturelle des pigeons*, Vol. 1, p. 7.
24 William Thomson [Baron Kelvin], *Popular Lectures and Addresses*, London, Macmillan, 1889, Vol. 1, p. 73.
25 Temminck, *Histoire naturelle des pigeons*, Vol. 1, p. 18.

CHAPTER VII

1 Hobsbawm, *The Age of Revolution*, p. 141.
2 Dunmore, *French Explorers in the Pacific*, Vol. 2, p. 228.
3 *Ibid.*
4 For a brief, but accurate account of the Verreaux family see J. J. Winterbottom, "Verreaux, Pierre Jules", *Dictionary of South African Biography*, 1972, Vol. 2, pp. 811–812.
5 From a notebook of Jules Verreaux, "Mammalogie et Ornithologie Australienne. 1844 et 1845", Bibliothèque du Muséum National d'Histoire naturelle, ms. 770, p. 304.
6 A useful bibliographical guide for the histories of these societies can be found in R. M. MacLeod, J. R. Friday, and C. Gregor, *The Corresponding Societies of the British Association for the Advancement of Science 1883–1929*, London, Mansell, 1975.
7 Rev. A. Hume, *The Learned Societies and Printing Clubs of the United Kingdom: Being an Account of Their Respective Origin, History, Objects, and Constitution* . . . , London, Longman, Brown, Green and Longmans, 1847, p. 13.
8 For example, one can find reference to the hiring of curators in Arthur Deane, *The Belfast Natural History and Philosophical Society. Centenary Volume 1821–1921*, Belfast, The Society, 1924, p. 15 (1834 curator hired); Goddard, *History of the Natural History Society of Northumberland, Durham and Newcastle Upon Tyne*, p. 50 (1835 curator hired); *Annual Report of the Council of the Shropshire and North Wales Natural History and Antiquarian Society, 1837*, p. 3 (1837 curator hired).
9 Great Britain, *Parliamentary Papers* (Session 223) (1845) (Bills: Public, Vol. 4), "An Act for Encouraging the Establishment of Museums in Large Towns", p. 441; which was amended by the "Public Libraries Act" of 1850 allowing any town council (not just towns of over 10,000) to establish museums or libraries: (Session 606) (1850) (Bills: Public, Vol. 7), p. 361.
10 *Reports of the Council of the Philosophical and Literary Society of Leeds*, 1869–70, **50**:4–5.
11 Edward Forbes, *On the Educational Uses of Museums*, London, Eyre & Spottiswoode, 1853, p. 14.
12 See Marjorie Plant, *The English Book Trade. An Economic History of*

the Making and Sale of Books, London, Allen & Unwin, 1965 (2nd edition), and Alfred Shorter, *Paper-Making in the British Isles. An Historical and Geographical Study*, Newton Abbot, David & Charles, 1971.

13 The importance of lithography was recognized immediately. See, for example, the review: "Swainson's Zoological Illustrations", *Edinburgh Philosophical Journal*, 1821, 4:209, and Hugh Strickland, "Report on the Recent Progress and Present State of Ornithology", *Report of the Fourteenth Meeting of the British Association for the Advancement of Science*, 1844: 202–203.

14 See Susan Pyenson, "Low Scientific Culture in London and Paris, 1820–1875", Ph. D., University of Pennsylvania, 1976, for a discussion of some of this literature.

15 "Introduction", *The Zoological Journal*, 1824, 1:iv.

16 Linnean Society of London, James E. Smith Correspondence, Vol. 23, pp. 98–99. The circulation of these early specialized journals was quite modest by modern standards. In his letter Konig refers to the *Annals of Botany* as having had a circulation of 250 and that it could have been 500. He was hoping for a circulation of as high as 1,000.

17 Among the more important natural history journals founded in the 1830's were: *Archiv für Naturgeschichte, Magasin de zoologie*, and *Magazine of Natural History* (n.s.).

18 British Museum, Add. ms. 37,188, fol. 303. Letter of April 8, 1834. Swainson elaborated on his position in his *A Preliminary Discourse on the Study of Natural History*, London, Longman, Rees, Orme, Brown, Green and Longman, 1834.

19 *Le Moniteur Universel*, 1841, no. 280, p. 2177.

20 See Baron Frédéric de La Fresnaye, *Essai d'une nouvelle manière de grouper les genres et les espèces de l'Ordre des Passereaux (Passeres L.) d'après leurs rapports de moeurs et d'habitation*, Falaise, Brée, 1838.

21 Linnean Soceity of London, Zoology Club mss., Kirby Letters, letter of September 12, 1822. This sentiment, the recognition of the need for a narrower more detailed focus, was a common theme in the years between 1820 and 1850. In the *Report from the Select Committee* it is often reiterated, and the suggestion for separate keepers, one for each department of natural history is repeatedly made.

22 For an interesting discussion of the difficulty of characterizing "professional" scientists see Susan Faye Cannon's *Science in Culture: The Early Victorian Period*, New York, Science History Publications, 1978, pp. 167–200.

23 Rachel Lauden, "Ideas and Organizations in British Geology: A Case Study in Institutional History", *Isis*, 1977, 68(244):527–538, questions the position that institutionalization is necessary for scientific development.

Since ornithology, *qua* discipline emerged before *ornithological* societies, journals, etc., the present study suggests a similar point.

24 [Anthelme Brillat-Savarin], *Physiologie du Gout, ou méditations de gastronomie transcendante; ouvrage théorique, historique et a l'ordre de jour, dédié aux Gastronomes parisiens, par un professeur*, Paris, Sautelet, 1826, Vol. 1, p. 142.

25 Rev. Leonard Jenyns, "Some Remarks on the Study of Zoology, and on the Present State of the Science", *Magazine of Zoology and Botany*, 1837, 1:1. This article is a shorter version of his address, of the same title, that he delivered to the British Association for the Advancement of Science in 1834.

26 *Ibid.*, p. 26.

27 "Prospectus de l'année 1833 (3e année)", *Magasin de zoologie*, 1833: 1–4.

28 "Principales pièces déposées au Ministre de L'Instruction Publique et relatives a la demande de souscription au Magasin de zoologie et a la Revue zoologique", *Revue et Magasin de Zoologie*, 1824, 1: vi–vii.

29 Hugh Strickland, "Report on the Recent Progress and Present State of Ornithology", *Report of the Fourteenth Meeting of the British Association for the Advancement of Science*, 1845:173.

30 *Ibid.*, p. 221.

31 British Museum (Natural History), Zoology Library, "Zoological Society of London Manuscript Reports of the Curator. 1836–1840". Report of May 15, 1839.

32 The quantity of literature on Audubon is enormous. For an introductory appreciation of his importance for the history of ornithology the following are useful: Anker, *Birds Books and Bird Art*, Ronsil, *L'art français*; Alice Ford, *John James Audubon*, Norman, University of Oklahoma Press, 1964; and Robert Henry Welker, *Birds and Men*, Cambridge, Mass., Harvard University Press, 1955.

33 Quoted in the informative biographical memoir, James Harley, "The Late Professor Macgillivray", *Report of the Council of the Leicester Literary and Philosophical Society*, 1853:105–164.

34 Frédéric Cuvier, "Ornithological Biography, or an Account of the habits of the birds . . . by John James Audubon", *Journal des savants*, 1833:706.

35 Frédéric Cuvier, "Nouveau recueil des planches coloriées d'oiseaux, pour servir de suite ou de complément aux planches enluminées de Buffon, par M. Temminck, conservateur du cabinet d'histoire naturelle de Leyde, et M. Meiffren Laugier, Baron de Chartrouse", *Journal des savants*, 1832:647.

36 Charles Lucien Bonaparte in his *A Geographical and Comparative List of the Birds of Europe and North America*, London, Voorst, 1838, p. vi, wrote "Throughout the list, I have quoted as Types of the Species under

consideration, the figures of the great works of Mr. Gould and M. Audubon on the Ornithology of the two regions, as they must be considered the standard works on the subject".

37 See Allen McEvey, *John Gould's Contribution to British Art*, Sydney, Sydney University Press, 1973. For an interesting discussion of Gould and the development of Lithography see Christine Jackson, *Bird Illustrators: Some Artists in Early Lithography*, London, Witherby, 1975.

38 See Sandra Herbert, "The Place of Man in the Development of Darwin's Theory of Transmutation. Part I. To July 1837", *Journal of the History of Biology*, 1974, 7(2):242—244.

39 Edward Blyth, "An Attempt to Classify the 'Varieties' of Animals, with Observations on the Marked Seasonal and Other Changes Which Naturally Take Place in Various British Species, and Which do not Constitute Varieties", *The Magazine of Natural History*, 1835, 3:40—53, and "Observations on the Various Seasonal and Other External Changes Which Regularly Take Place in Birds, More Particularly in Those Which Occur in Britain, with Remarks on Their Great Importance in Indicating the True Affinities of Species; and upon the Natural System of Arrangement", *The Magazine of Natural History*, 1836, 9:393—409 and 505—514.

40 Hugh Strickland developed his ideas in "On the True Method of Discovering the Natural System in Zoology and Botany", *The Annals and Magazine of Natural History*, 1814, 6:184—194 and in "On the Structural Relations of Organized Beings", read before the Ashmolean Society of Oxford, March 10, 1845, and printed in William Jardine, *Memoirs of Hugh Strickland, M.A.*, London, Voorst, 1858, pp. 348—356.

41 Strickland, "On the True Method", p. 190.

42 *Ibid.*

43 *Ibid.*

44 Isidore Geoffroy Saint-Hilaire, "Considérations sur les caractères employés en ornithologie pour la distinction des genres, des familles et des ordres, et détermination de plusieurs genres nouveaux", *Nouvelles Annales du Muséum d'Histoire Naturelle*, 1832, 1:357—397.

45 N. A. Vigors and Thomas Horsfield, "A Description of the Australian Birds in the Collection of the Linnean Society; with an Attempt at Arranging Them According to Their Natural Affinities", *Transactions of the Linnean Society of London*, 1827, 15:170—172 tried to argue the relationship between the Linnean and quinary systems. Swainson in his numerous publications argued for the relationship between the quinary system and a form of Natural Theology.

46 William Swainson, *A Treatise on the Geography and Classification of Animals*, London, Longman, Rees, Orme, Brown, Green and Longman, 1835, p. 242.

[47] *Ibid.*, p. 245.

[48] Strickland, "Report on the Recent Progress", p. 177.

[49] See Johann Jakob Kaup, *Classification der Saugethiere und Vogel*, Darmstadt, Leske, 1844. Lenoir makes a useful distinction between the *Naturphilosophie* of Oken *et al.* and Blumenbach and his followers. See his "The Göttingen School".

[50] Kaup, for example, wrote to George Robert Gray that "Your published works on Ornithology we consider in Germany to be the best which we have and many of our great Museums are arranged after your System". British Museum, Egerton ms. 2348, fol. 218, undated letter from 1851.

[51] "Report of a Committee Appointed 'to Consider of the Rules by Which the Nomenclature of Zoology May be Established on a Uniform and Permanent Basis'", *Report of the Twelfth Meeting of the British Association for the Advancement of Science*, 1842:106–107.

[52] *Ibid.*, pp. 107–108.

[53] See L. Elie de Beaumount, *Notice sur les travaux scientifiques de son altesse le Prince Charles-Lucien Bonaparte*, Paris, Bénard, 1866; Maurizia Capelletti Alippi, "Bonaparte, Carlo Luciano, principe di Canino", *Dizionario Biografico Degli Italiani*, 1969, 11:549–556; and Erwin Stresemann, *Die Entwicklung*, pp. 155–171.

[54] The Bibliothèque de Muséum National d'Histoire naturelle possesses the extensive scientific correspondence of Bonaparte, as well as an enormous collection of his scientific papers. They contain ample evidence of his position in the scientific community.

[55] "Report from the Select Committee", Vol. 2, 1836, p. 45.

[56] Bibliothèque du Muséum National d'Histoire naturelle, ms. 119.

CHAPTER VIII

[1] Charles Lucien Bonaparte, *Conspectus generum avium*, Leyden, Brill, 1850–1857, vol. 1, p. i.

[2] Foucault, *Les mots et les choses*, p. 139. I have quoted from the English translation, Michael Foucault, *The Order of Things. An Archeology of the Human Sciences*, London, Tavistock, 1970, pp. 127–128. It should be noted that Foucaut's ideas have been in flux. I am describing his ideas as expressed in *Les mots et les choses*. This choice is not arbitrary, but rather it is due to the continuing influence that book has had on historians. For a recent example see Stephen Cross, "John Hunter, the Animal Oeconomy, and Late Eighteenth-Century Physiological Discourse", *Studies in History of Biology*, 1981, 5:1–110. Foucault's characterization of the changes in natural history is very broad and contains elements of both of the interpretations of the

history of natural history referred to in the Introduction (*i.e.*, a replacement of natural history by "biology", and a new temporal dimension in the conception of nature).

3 Foucault, *The Order of Things*, p. 74.

4 *Ibid.*, p. 217.

5 See my "Research Traditions in Eighteenth-Century Natural History".

6 The history of the emergence of comparative anatomy as a scientific discipline has yet to be written. Some elements of the story are in Bernard Balan, *L'Ordre et le temps. L'Anatomie Comparée et l'histoire des vivants au XIXe siècle*, Paris, Vrin, 1979.

7 Daubenton was Buffon's collaborator on the first section of the *Histoire naturelle*. He did not, however, provide anatomical studies of each species of bird for the *Histoire naturelle des oiseaux* as he had for each of the species of the quadrupeds in the preceding section of the *Histoire naturelle*. See my "Buffon and Daubenton: Divergent Traditions within the *Histoire naturelle*", *Isis*, 1975, 66(231):63–74.

8 Coleman, *Biology in the Nineteenth Century*, p. 3.

9 Lynn Barber, *The Heyday of Natural History 1820–1870*, London, Jonathan Cape, 1980 raises the issue but only superficially treats it.

10 A good introduction to the literature on this subject is in the notes to Nathan Reingold, "Definitions and Speculations: the Professionalization of Science in America in the Nineteenth Century", in Alexandra Oleson and Sanborn C. Grown (eds.), *The Pursuit of Knowledge in the Early American Republic. American Scientific and Learned Societies from Colonial Times to the Civil War*, Baltimore, The Johns Hopkins University Press, 1976, pp. 33–69.

11 *Ibid.*

12 Reingold has often cautioned against oversimplification of the historical record. See, for example, his article, "National Aspirations and Local Purposes", *Transactions of the Kansas Academy of Science*, 1968, 71(3): 235–246.

13 *Ibid.*, p. 236.

14 Maurice Crosland, "The Development of a Professional Career in Science in France", in Maurice Crosland (ed.), *The Emergence of Science in Western Europe*, New York, Science History Publications, 1976, pp. 154–155.

15 Joseph Ben-David, *The Scientist's Role in Society. A Comparative Study*, Englewood Cliffs, Prentice-Hall, 1971.

16 Roger Hahn suggests how complicated an understanding of the story is for France in his "Scientific Careers in Eighteenth-Century France", in Maurice Crosland (ed.), *The Emergence of Science in Western Europe*, New York, Science History Publications, 1976, pp. 127–138. Also see Robert Fox, "Scientific Enterprise and the Patronage of Research in France

1800–1870", *Minerva*, 1973, **11**(4):442–473, and Dorinda Outram, "Politics and Vocation: French Science, 1793–1830", *British Journal for the History of Science*, 1980, **13**:27–43.

[17] W. J. Reader, *Professional Men. The Rise of the Professional Classes in Nineteenth-Century England*, London, Weidenfeld and Nicholson, 1966, p. 147.

[18] See George Daniels, "The Process of Professionalization in American Science: the Emergent Period, 1820–1860", *Isis*, 1967, **58**(192): 151–166, and Geoffrey Millerson, *The Qualifying Associations. A Study in Professionalization*, London, Routledge & Kegan Paul, 1964.

[19] See, for example, Marianne Gosztonyi Ainley, "The Contribution of the Amateur to North American Ornithology: a Historical Perspective", *The Living Bird*, 1979–80, **18**:161–177, and Allen, *The Naturalist in Britain*.

[20] See Susan Faye Cannon, *Science in Culture*.

[21] Roy Porter, "Gentlemen and Geology: the Emergence of a Scientific Career, 1660–1920", *The Historical Journal*, 1978, **21**(4):810. Unfortunately Porter confines himself to British geologists, and it is not clear that one can generalize from it.

[22] See quotation cited in chapter seven.

[23] Stephen Toulmin gives a detailed discussion on the philosophic differences in his *Human Understanding*, Princeton, Princeton University, 1972.

[24] Everett Mendelsohn, "The Emergence of Science as a Profession in Nineteenth-Century Europe", in Karl Hill (ed.), *The Management of Scientists*, Boston, Beacon Press, 1964, pp. 40–41.

[25] Levaillant, *Histoire naturelle des Perroquets*, Vol. 1, p. i.

[26] William MacLeay, *Horae Entomologicae; or Essays on the Annulose Animals*, London, Bagster, 1819, p. vi.

[27] Jenyns, "Some Remarks on the Study of Zoology", (1839), p. 26.

[28] See Sheets-Pyenson, "War and Peace in Natural History Publishing", pp. 71–72. Jardine did a total of fifteen of the entire forty volume set.

[29] [Isidore de Salles], *Histoire naturelle drolatique et philosophique des Professeurs du Jardin des plantes, des aides-naturalistes, préparateurs, etc., attachés à cet établissement, accompagnée d'épisodes scientifiques et pittoresques, par Isid. S. de Gosse. Avec des annotations de Frédérick Gérard*, Paris, Sandré, 1847, p. 150. Berthold Schwarz was an alleged inventor of gunpowder in the middle ages. Robert Macaire was a villain in a popular melodrama of the period, and was the prototype for a series of lithographs by Honoré Daumier which depicted thievery of various sorts.

[30] Buffon, *HNO*, Vol. 2, p. 523.

[31] George Edwards, *A Natural History of Birds*, London, Printed for the author, 1743–1751, Vol. 4, p. ii.

[32] See Cannon, *Science and Culture*, p. 3.

[33] See Cannon, *Science and Culture* which also contains valuable bibliographical information. David Hull has attempted to give a philosophical analysis of the Victorian philosophy of science in his *Darwin and His Critics*, Cambridge, Mass., Harvard University Press, 1973.

[34] See, for example, John Frederick Herschel, *A Preliminary Discourse on the Study of Natural Philosophy*, London, Longman, Rees, Orme, Brown, & Green, 1830.

[35] "Introduction", *Annales des sciences naturelles*, 1824, 1:ix.

[36] Neville Wood, *The Ornithologist's Text-Book. Being Reviews of Ornithological Works; with an Appendix, Containing Discussions on Various Topics of Interest*, London, John Parker, 1836, p. 153.

[37] Herbert, "The Place of Man in the Development of Darwin's Theory", p. 244.

[38] David Kohn, "Theories to Work By: Rejected Theories, Reproduction, and Darwin's Path to Natural Selection", *Studies in History of Biology*, 1980, 4:73.

[39] Francis Darwin (ed), *The Life and Letters of Charles Darwin*, New York, Appleton and Co., 1896, Vol. 1, pp. 315–316.

[40] There have been some excellent studies of Darwin's reception in individual countries, however, not much has been written of a comparative nature except for Thomas Glick (ed.), *The Comparative Reception of Darwinism*, Austin, University of Texas Press, 1974. Yvette Conry in her *L'introduction du Darwinisme en France au XIXe siècle*, Paris, Vrin, 1974 does a good job of examining the lack of response to Darwin in France.

[41] Newton, *A Dictionary of Birds*, p. 79.

[42] See Conry, *L'introduction du Darwinisme* and Joseph Schiller, *Claude Bernard et les problèmes scientifiques de son temps*, Paris, Les Editions du Cèdre, 1967.

[43] Richard French has shown a subtle effect on British physiology in his "Darwin and the Physiologists, or the Medusa and Modern Cardiology", *Journal of the History of Biology*, 1970, 3(2):253–274.

[44] A notable exception is Lucile Brockway, *Science and Colonial Expansion. The Role of the British Royal Botanic Gardens*, New York, Academic Press, 1979.

[45] *Ibid.*, p. 39. For a good description of the biological significance of the early contacts of Europeans with the New World see Alfred W. Crosby, Jr., *The Columbian Exchange. Biological and Cultural Consequences of 1492*, Westport, Conn., Greenwood Press, 1972.

[46] Morris Berman in *Social Change and Scientific Organization* discusses the significance of the discovery in the context of gaining support for science in nineteenth-century Britain.

[47] See S. Peter Dance, "Hugh Cuming (1791–1865) Prince of Collectors",

Journal of the Society for the Bibliography of Natural History, 1980, **9**(4): 477–501.

[48] For an interesting discussion of the complexity of these changes and their relationship to the social and economic events of the period see Fritz Ringer, *Education and Society in Modern Europe*, Bloomington, Indiana University Press, 1979.

[49] See Joseph Fayet, *La Révolution française et la science*, Paris, Marcel Rivière, 1960, pp. 110–119.

[50] Allen, *The Naturalist in Britain*, has a good discussion on this subject for the British example.

[51] *Ibid.*, p. 74.

[52] *Ibid.*, p. 75.

[53] Charles Babbage, *On the Economy of Machinery and Manufactures*, 4th ed., London, Charles Knight, 1835, p. 386.

[54] See Arnold Thackray, "Natural Knowledge in Cultural Context: the Manchester Model", *The American Historical Review*, 1974, **79**(3):672–709.

[55] *Ibid.*, pp. 674–675.

[56] *Ibid.*, p. 693.

[57] George Johnston, "Address to the Members of the Berwickshire Naturalists' Club", *History of the Berwickshire Naturalists' Club*, 1834, p. 11.

[58] Babbage, *On the Economy of Machinery and Manufactures*, p. 379.

[59] "The Edinburgh Journal of Natural and Geographical Science. New Series", n.p., n.p., n.d., p. 1.

[60] *Ibid.*, p. 2.

[61] See Phyllis Deane, *The First Industrial Revolution*, Cambridge, Cambridge University Press, 1965.

INDEX

Adanson, Michel 9, 18
Affinities 110–11
African birds 18, 32–9, 63, 87. *See also*
exotics; Local faunas, Exotic
Agronomy 1–2, 47
Albin, Eleazer 4
Aldrovandi, Ulisse 5, 21
Allen, David xx, 152–3
Anatomy 13, 31, 47
Arsenical soap 54
Argenville, Antoine-Joseph Desallier d'
49–50
Artur, François 9, 18
Asiatic Society 45
Aubry, abbé 10
Audebert, Jean-Baptiste 86–7
Audubon, John James 42, 104–7
Audience for ornithology 14, 74, 77,
97, 99–100
Autie, Léonard 4
Aviaries 2
Azara, Félix 45

Babbage, Charles 98, 153–4, 156
Baillon, M. 18
Bandeville, Mme la présidente 10
Banks, Joseph 71, 148
Barraband, Jacques 87
Barrière, Pierre 4
Barrington, Daines 28
Baudin, Nicolas 40, 53
Bechstein, Johann Matthaeus 30, 77
Bécoeur, Jean-Baptiste 54, 65
Beechey, Frederick William 41
Behavior, bird 8, 22, 85, 88, 93, 99, 106
Belon, Pierre 5
Ben-David, Joseph 129
Bentinck, Count 9
Bernard, Claude xvii
Bewick, Thomas 28–9, 77
Bexon, abbé Gabriel-Leopold 21

Biological science, relationship to general
culture xvi, 128, 137, 147–157
Biology: concept of xvii 123; early use
of word xvi–xvii, 126
Birds: commercial collecting 38–9;
complete bird list 115–9, 122; field
collecting 33–46, 60–1, 93–4, 141,
148–9; knowledge of in the eight-
eenth century 1–6; illustrations of
12, 86–8; in cooking, 1–2; in decorat-
ive arts 2; in fashion 2–4; in heraldry
1; place in amusement 2; private
collections of 8–10, 35, 49–67.
See also Museums
Blainville, Henri-Marie Ducrotay de 83,
109
Blumenbach, Johann Friedrich 84, 111
Blyth, Edward 104, 108
Boie, Heinrich 38
Bonaparte, Charles Lucien 59, 104,
116–9, 121–2
Bonelli, Franco Andrea 32, 62
Borde, M. de la 18
Boubier, Maurice xx
Bougainville, Hyacinthe 41
Brehm, Christian Ludwig 31–2, 104,
109
Brillat-Savarin, Anthelme 101
Brisson, Mathurin-Jacques 5, 7–15, 26,
68–9, 116, 121, 124; influence of
70, 127
British Museum 45, 54, 57–8, 67, 71,
116, 118
Bruce, James 18
Buchanan, Francis 44
Buffon, Georges-Louis Leclerc de Buffon
5, 7, 15–26, 52, 68–70, 76, 121, 124;
influence of 70, 127
Bullock, William 51–2, 55–7
Burchell, William 34–5, 46

187